Lecture Notes in Mobility

Series editor

Gereon Meyer, Berlin, Germany

More information about this series at http://www.springer.com/series/11573

Gereon Meyer · Sven Beiker
Editors

Road Vehicle Automation 3

Springer

Editors
Gereon Meyer
VDI/VDE Innovation + Technik GmbH
Berlin
Germany

Sven Beiker
Stanford University
Palo Alto, CA
USA

ISSN 2196-5544 ISSN 2196-5552 (electronic)
Lecture Notes in Mobility
ISBN 978-3-319-40502-5 ISBN 978-3-319-40503-2 (eBook)
DOI 10.1007/978-3-319-40503-2

Library of Congress Control Number: 2014941235

Printed on acid-free paper

This Springer imprint is published by Springer Nature
The registered company is Springer International Publishing AG Switzerland

Preface

Higher-level automated driving keeps moving at full speed toward implementation: Universities are researching solutions for environmental perception, vehicle localization, and human interaction. Companies develop new vehicle and service concepts for automated driving and start providing automated driving functions via over the air software upgrades. Governments are working on the necessary regulatory frameworks; for instance, just very recently, NHTSA decided to interpret software as the "driver" of a self-driving car.

Still, many challenges remain and require more work, for instance, automated driving in mixed traffic, the security of sensitive vehicle data, the need for a common language and for a joint strategy of the automotive and IT sectors. However, now is the time for extended testing and piloting of high-level automated driving under real-time conditions, particularly in the complexity of an urban environment, or across borders, and in consideration of multiple vehicle types.

In this context, it is a great honor for us to edit the Road Vehicle Automation books that are published under the umbrella of the Springer series Lecture Notes in Mobility. The book at hand is the third volume. It summarizes the lively discussions on the political, behavioral, technical, and organizational issues of automated driving that took place at the Automated Vehicles Symposium (AVS) 2015 in Ann Arbor, Michigan (USA). Many speakers and breakout session organizers kindly contributed chapters to this book, for which we would like to thank them sincerely on behalf of all readers. This gives researchers, engineers, and decision-makers a unique opportunity to refer to presentations and discussions after the conference, carry on their own work, and educate others on one of the most transformational transportation trends of our times.

We are proud to say that, thanks to the high quality of the contributions, the comprehensiveness of the topics covered, and the availability on the Internet and in university libraries worldwide, the Road Vehicle Automation books are having quite an impact on the current expert's discussions on automated driving. The fact that as of today, Road Vehicle Automation 1 and 2 have gotten already 35 and 10 thousand downloads, respectively, documents the relevance of this series in an impressive way. This success motivates us to further continue the series.

We are particularly grateful to the organizers of the AVS 2015, the Transportation Research Board (TRB), and the Association for Unmanned Vehicle Systems International (AUVSI), for their kind support. Special thanks go to Jane Lappin, Steve Shladover, and Bob Denaro from TRB for working with us on this publication. Furthermore, we would like to thank Jan-Philip Schmidt from Springer and Jakob Michelmann from VDI/VDE-IT for all their support during the editorial processes.

Berlin, Germany
Palo Alto, USA
May 2016

Gereon Meyer
Sven Beiker

Contents

Introduction: The Automated Vehicles Symposium 2015

Steven E. Shladover, Jane Lappin and Robert P. Denaro

Abstract The 2015 Automated Vehicles Symposium built on the successes of the predecessor meetings, with an even larger and more diverse roster of participants and a broader selection of breakout sessions. It was organized in cooperation with the University of Michigan's Mobility Transformation Center, which provided an opportunity for the Symposium participants to experience a wide range of vehicle demonstrations at their new MCity test site. The plenary and poster presentations and breakout discussions continued to provide the meeting participants with the most up-to-date and authoritative information about the current international state of development of road vehicle automation systems, making this the essential meeting for industry, government and research people interested in the subject.

Keywords Road vehicle automation · Road transport automation · Automated vehicles · Autonomous vehicles · Self-driving vehicles

S.E. Shladover (✉)
University of California PATH Program, 1357 South 46th Street,
Building 452, Richmond, CA 94804, USA
e-mail: steve@path.berkeley.edu

J. Lappin
Volpe National Transportation Systems Center, 55 Broadway,
Cambridge, MA 02142, USA
e-mail: Jane.Lappin@dot.gov

R.P. Denaro
ITS Consultant and Advisory Board Member of Motus Ventures,
PO Box 1587, Grand Lake, CO 80447, USA
e-mail: bobdenaro@aol.com

G. Meyer and S. Beiker (eds.), *Road Vehicle Automation 3*,
Lecture Notes in Mobility, DOI 10.1007/978-3-319-40503-2_1

1 Overview

The 2015 Automated Vehicles Symposium was organized and produced through a partnership between the National Academies of Science and Engineering Transportation Research Board (TRB) and the Association for Unmanned Vehicle Systems International (AUVSI), continuing the pattern established by the 2014 Symposium. This meeting was organized to serve their constituencies' interests in understanding the impacts, benefits, challenges and risks associated with increasingly automated road vehicles and the environments in which they operate. It brought together key government, industry and academic experts from around the world with the goal of identifying opportunities and challenges and advancing automated vehicle (AV) research across a range of disciplines.

The symposium took place over 5 days, 20–24 July, with 3 days of core activities and ancillary sessions on the first and last days. The morning plenary sessions included presentations from the public sector, automakers and suppliers and research institutes and the afternoons were devoted to seventeen breakout sessions for deeper investigation and discussion of selected topics. Receptions and poster sessions followed the close of the breakout sessions on Tuesday and Wednesday afternoons.

The breakout sessions were each organized by committees of volunteers to address a wide range of topics. Four of the breakout sessions spanned both afternoons of the Symposium, providing more time for exploration in greater depth and breadth:

- Beyond Single Occupancy Vehicles: Automating Transit and Shared Mobility
- Human Factors in the Design of Road Vehicle Automation
- Legal Aspects of Automated Vehicles, including Liability, Insurance and Ethics
- Early Deployment Opportunities for Connected Automation Systems.

The other thirteen breakout sessions covered a single afternoon each:

- Energy and Demand
- Physical and Digital Infrastructure
- Prioritizing Public Policy Challenges for Automated Vehicles
- Truck Automation
- Wireless Connectivity for Automated Vehicles
- Cybersecurity for Automated Vehicles
- Implications of Automated Vehicles for Planning
- Integrated Traffic Flow Models and Analysis for Automated Vehicles
- Traffic Signal Control with Connected and Automated Vehicles
- Vulnerable Road Users—How Can AVs Help Keep them Safe and Mobile?
- Impact of Connected and Automated Vehicles on Traffic Management Systems and Operational Strategies
- Verification and Validation of On-Road Automated Vehicles
- Enabling Technologies for Road Vehicle Automation.

The symposium also involved several related meetings that occurred before and following the main meeting:

- Opening ceremony for the University of Michigan's Mobility Transformation Center
- AASHTO Automated Vehicle Public Policy Workshop
- U.S. DOT Listening Session
- TRB Workshop on Envisioning Automated Vehicles within the Built Environment: 2020, 2035 and 2050.
- Meetings of the TRB Automated Transit Systems Committee and Freeway Operations Committee and its subcommittees
- U.S.–Japan–EU Trilateral Working Group on Automation in Road Transportation.

In keeping with TRB practice, the plenary and breakout sessions were planned and produced by volunteers whose expertise and interests informed the content of the sessions. In keeping with AUVSI practice, the production of the symposium was professionally managed by dedicated conference and logistics managers. The AVS15 Executive Committee reflected this mix of the two organizations:

David Agnew, Continental Automotive, Member, AUVSI Board of Directors; Richard Bishop, AUVSI subject matter expert on automation; Richard Cunard, Senior Program Officer, Traffic and Operations Engineer, TRB; Bob Denaro, ITS Consultant, Chair, TRB Joint Subcommittee on the Challenges and Opportunities for Road Vehicle Automation; Jane Lappin, Volpe National Transportation Systems Center, Chair, TRB Intelligent Transportation Systems Committee (AHB15); James Misener, Qualcomm, Symposium Demonstrations Coordinator; Steven Shladover, University of California PATH Program, Chair, TRB Vehicle-Highway Automation Committee (AHB30); John Maddox, Assistant Director, University of Michigan Mobility Transformation Center and Director of Collaborative Strategies, University of Michigan Transportation Research Institute; Brian Wynne, President and CEO, AUVSI; Lindsay Voss, Senior Program Development Manager, AUVSI.

2 Symposium Attendees

More than 870 registrants participated in the symposium. Attendees represented a wide range of organizations from government and industry to the academic-, public-, and private-sector research communities. One of the strengths of the meeting was the breadth of interests represented, with 56.5 % from industry, 18 % from public agencies and 25.5 % from academic and research institutions. This represented a higher proportion from industry and a lower proportion from academic and research institutions than at AVS14. The industry participants included 86 people from automotive OEMs and 68 from suppliers.

These participants represented disciplines ranging from engineering to psychology to law. Twenty-four countries (with 156 participants from outside the U.S.) and forty U.S. states were represented among the meeting participants. The largest delegation from outside the U.S. came from Japan, with 46 participants. Michigan, as the host state, had the largest number of attendees from within the U.S., followed by California and the national capital region (DC, Maryland, and Virginia).

3 Demonstrations

The scheduling of AVS15 to coincide with the opening of the Mobility Transformation Center (MTC) made it possible to schedule vehicle demonstrations in coordination with the MTC event. Most of these demonstrations were given at the MTC site, with shuttle bus service to facilitate movements of visitors between the conference hotel and the MTC. The demonstrators included Bosch, which demonstrated automatic emergency braking for pedestrians and a lane-keeping and adaptive cruise control combination called "highway assist". Delphi demonstrated the Audi SQ5, the car that drove across the country, and a Jaguar F-Type equipped with Driver State Sensing. Xerox demonstrated its Vehicle Passenger Detection System™ which enables transportation agencies or law enforcement to monitor lane use, CloudParc parking monitoring system, and MERGE™ for parking location.

4 Keynote Talks

Dr. Mark Rosekind, the Administrator of the National Highway Traffic Safety Administration (NHTSA) gave the opening plenary address, discussing his agency's concerns about traffic safety and the potential for automated vehicles to improve traffic safety with a fundamentally new goal of preventing crashes before they occur. He noted USDOT Secretary Anthony Foxx's emphasis on the importance of technology innovation for transportation.

Dr. Chris Urmson, the Director of the Google X Self-Driving Car Program, gave the second keynote address, describing the motivations behind the Google program and noting that the annual fatality rate on the Nation's highways is roughly the equivalent of a 737 airliner crash every business day. He showed videos to demonstrate the progress Google has been making on the automation of driving in complicated urban scenarios with further nuanced behavior such as their automated vehicles programmed to maneuver out of other vehicles' blind spots. He also noted Google's designed-in redundancy of actuators and sensors with multiple LiDARs, cameras and radars.

5 Plenary Panel Sessions

Vehicle Manufacturers and Suppliers:

- Bob Denaro, Moderator
- Michael Pozsar, Vice President, Electronic Controls, Electronics and Safety, Delphi
- Dr. Kay Stepper, Vice President, Head of Regional Business Unit Driver Assistance and Automated Driving, Robert Bosch LLC
- Dr. Cem U. Saraydar, Director, Electrical and Controls Systems Research Lab, General Motors.

Identifying and Addressing Key Research Questions:

- Dr. Steven Shladover, Moderator
- Legal Issues Addressed in the EU Adaptive Project: Andreas Knapp, Daimler AG
- Key Safety Principles for Automation by Automation Level: Levasseur Tellis, Technical Specialist, Functional Safety, Ford Motor Co.
- Michigan Mobility Transformation Center Research Roadmap: John Maddox, Assistant Director, Michigan Mobility Transformation Center
- Ethical Considerations for Vehicle Automation Systems: Dr. Chris Gerdes, Professor, Mechanical Engineering, Stanford University
- EU-US Collaboration in Road Transport Automation: Dr. Peter Sweatman, Director and Research Scientist, UMTRI.

Private Investment in Vehicle Automation:

- John Casesa, Vice President of Global Strategy, Ford Motor Co., Moderator
- Philipp von Hagen, Member of Executive Board, Porsche Automobil Holding SE
- Zach Barasz, Kleiner Perkins Caufield and Byers
- Glenn Mercer, Industry Analyst
- Chris Thomas, Founder and Partner, Fontinalis.

Automated Vehicle Verification:

- John Maddox, Assistant Director, Mobility Transformation Center, Moderator
- Stephanie Dougherty, Chief of Enterprise Planning and Performance, California Department of Motor Vehicles
- Ibro Muharemovic, Continental Automotive Systems Inc.
- Felix Fahrenkrog, Manager Active Safety ADAS, Driver Assistance, RWTH Aachen University.

State and City Level Issues:

- Jane Lappin, Moderator
- Dr. Johanna Zmud, Senior Research Scientist, Texas A&M Transportation Institute
- Paul Steinman, District Secretary, Florida Department of Transportation
- Leon Daniels, Managing Director, Surface Transport, Transport for London.

6 Plenary Presentation Sessions

International Automated Vehicle Initiatives:

- CityMobil2 Project—Dr. Adriano Alessandrini, University di Roma La Sapienza
- The Pathway to Automation and Connectivity—Meaningful Momentum in the UK: Michael Hurwitz, Director, Energy, Technology and International, UK Department for Transport
- Drive Sweden: Jan Hellaker, Head of Automation, Lindholmen Science Park AB
- French National Automated Vehicle Program: Dr. Christian Rousseau, Renault
- World Economic Forum: Alex Mitchell, Director, Head of Automotive Industry, World Economic Forum.

User Considerations:

- Automated Vehicles and Public Perception: Kristin Kolodge, Executive Director, Driver Interaction, J.D. Power
- Human Factors Evaluation of Level 2 and Level 3 Automated Driving Concepts: Dr. Myra Blanco, Research Scientist, Virginia Tech Transportation Institute
- Automated Vehicles and Human Factors: Patrice Reilhac, Innovation and Collaborative Research Director, Comfort and Driving Assistance Business Group, Valeo
- Automated Trucking Applications: Dr. Josh Switkes, Founder and CEO, Peloton Technology Inc.
- Digital Infrastructure: Ogi Redzic, Senior Vice President, Automotive, HERE.

Public Agency Automated Vehicle Initiatives:

- Japan: Hajime Amano, President, ITS Japan
- European Commission: Ludger Rogge, Research Programme Officer, DG Research and Innovation, European Commission
- U.S. Department of Transportation: Kevin Dopart, Program Manager, Connected Vehicle Safety and Automation, Intelligent Transportation Systems Joint Program Office
- National Highway Traffic Safety Administration: Nathaniel Beuse, Associate Administrator, Vehicle Safety Research, National Highway Traffic Safety Administration
- U.S. Department of Energy: Reuben Sarkar, Deputy Assistant Secretary for Transportation, Energy Efficiency and Renewable Energy, U.S. Department of Energy.

7 Breakout Sessions

The breakout sessions provided opportunities for more in-depth consideration of specific topic areas among groups of people with focused interests in those areas. With smaller groups, they could be more interactive than the large plenary sessions, with ample opportunities for questions and answers and debates. The primary findings from the breakout discussions were reported back to the plenary group on the final morning of the Symposium, in four panels based on thematic groupings.

7.1 Infrastructure and Operations Breakout Sessions

Early Deployment Opportunities for Connected Automation Systems

This group focused on Level 1 automation systems, with the earliest opportunities for deployment, beginning with systems that are already commercially available. Roadway operators need to be receptive to supporting field testing, including facilitating use in managed lanes, but it is not clear whether they perceive sufficient benefits yet to justify investment in infrastructure changes, such as improving roadway markings. They need to see validated benefits estimates and have their concerns about cyber security threats allayed.

Impact of Connected and Automated Vehicles on Traffic Management Systems and Operational Strategies

This group identified the need for outreach to practitioners at city and county levels so that they can be better integrated into the discussions about automation and better prepared to respond to it. They are eager to understand how the automated vehicle market will grow and when it could potentially enable the elimination of some of today's traffic management infrastructure, such as detectors. They would like to use newly available data from automated vehicles to improve traffic management procedures.

Integrated Traffic Flow Models and Analysis for Automated Vehicles

This group focused on identifying research needs and reconciling divergent goals of vehicle manufacturers, users and roadway infrastructure operators. They see the need for improved data about the interactions between drivers of conventional vehicles and automated vehicles and potential driver behavior changes that will have to be represented in improved models, and will need to use improved models to predict the effects of various market penetrations of vehicle automation on highway bottlenecks, merges and signalized intersections.

Traffic Signal Control with Connected and Automated Vehicles

The group concluded that traffic signals will still be needed 20 years from now, even as more automation capabilities become available. Key challenges are expected to involve understanding the interactions between human drivers and automated vehicles at intersections, and determining what types of new behaviors emerge from those interactions.

Physical and Digital Infrastructure

This group had separate discussions of the physical and digital infrastructure issues, since they perceived the issues to be sufficiently different. They realized that any conclusions they draw would be significantly dependent on specific operational scenarios, particularly when considering the difference between roadway facilities dedicated to automated vehicles and facilities handling mixed traffic. They see a role for government in facilitating development of common standards for digital infrastructure.

7.2 *Technology Breakout Sessions*

Enabling Technologies for Road Vehicle Automation

This group discussed the technology trends expected for the next 3–5 years and the gaps in available technologies. They identified the lack of a common clearinghouse for information about technology capabilities as a problem, and concluded that people working on vehicle technology and infrastructure technology need to be sharing more of their knowledge. They also discussed the desirability of a free and open core set of data from operating vehicles.

Wireless Connectivity for Automated Vehicles

It was generally agreed that wireless connectivity and cooperative ITS are going to be necessary to support automation, but there were differences of opinion about the relative merits of DSRC and cellular LTE technologies for meeting automation system needs.

Cybersecurity for Automated Vehicles

This group recognized that security for automated vehicles will evolve from the current security considerations for connected vehicles, without any dramatic jump or transition. They considered cloud connectivity to be essential to automated vehicle operation, hence the need for cybersecurity. They identified the lack of a security threat catalog and of automotive security standards as an important gap, but also questioned the feasibility of defining such standards. Controversies included questions about what infrastructure support is needed, whether a secure public key infrastructure (PKI) is possible, and how to reconcile security, privacy and safety goals.

Verification and Validation of On-Road Automated Vehicles

Verification and validation were discussed in the context of a broader system engineering paradigm for vehicle system development, based on definition of requirements. Current challenges were identified to include the lack of standards and commonly accepted terminology, the lack of commonly accepted expectations of performance and safety and the high cost and complexity of the existing approaches to verification and validation.

7.3 User-Oriented Breakout Sessions

Human Factors in the Design of Road Vehicle Automation

The human factors breakout group focused on identifying the most important research needs in their field, based on the recognition that if the human factors issues are not addressed effectively by the early automated vehicles there is likely to be a public backlash against automation. Important questions were identified as: What feedback should an automated vehicle give to the driver (including whether it should inform the driver about its level of confidence in its current actions)? How should the automated vehicle monitor and respond to its driver? How should the system tell the driver that it is reaching the boundary of its operational design domain, where the driver needs to take over? How should the driver be alerted to regain his or her attention? The group noted that in general the necessary symbiotic relationship between the vehicle and the person is as yet poorly understood.

Vulnerable Road Users—How Can AVs Help Keep them Safe and Mobile?

The vulnerable road user group was concerned about how limited road space will be allocated among competing users in the future, particularly if the population of automated vehicles grows significantly. How will this impact the mobility of seniors, teens, pedestrians, bicyclists and the disabled? Will they be safer if automated vehicles follow traffic rules more strictly than drivers do today?

Beyond Single Occupancy Vehicles—Automating Transit and Shared Mobility

The focus of this group was on user needs, and whether those would be better served by automating transit vehicles or changing to a shared-use fleet of automated vehicles. The group was relatively evenly divided between these alternatives, but agreed that transit needs to be emphasized early in the development of road vehicle automation in order to deter dramatic growth in vehicle miles traveled.

Truck Automation

This group focused on the challenges that need to be overcome for truck automation to realize its promise as an early deployment opportunity for road vehicle automation. Key challenges were identified to be: development of standard testing and evaluation guidelines and procedures, harmonization of state regulations for long-distance trucking applications and the need to convincingly demonstrate the safety and acceptability (to truck drivers and the general public) of close-formation platooning of trucks.

7.4 Policy and Planning Breakout Sessions

Prioritizing Public Policy Challenges for Automated Vehicles

Primary themes in this breakout were the large number of uncertainties surrounding the impact of road vehicle automation and the need to educate the public and policy makers about what is actually known and unknown. Credible research and demonstrations are needed to reduce the uncertainties and to narrow the very wide range of current reactions to automation within the political world.

Legal Aspects of Automated Vehicles, Including Liability, Insurance and Ethics

This group tackled a wide range of legal issues. They concluded that the existing insurance and liability legal structures are adequate to manage the uncertainties about automation, but international harmonization of laws will be needed. They encountered some controversy in discussions about learning by doing to initiate deployment and about modernization of the Vienna and Geneva Conventions on Road Traffic, but also concluded that some of the data privacy issues are not well covered under existing laws.

Implications of Automated Vehicles for Planning

The planning group was concerned about the need to better predict an uncertain future given their work on 25-year plans and the very long time scales for development or modernization of public infrastructure. They need more clearly articulated descriptions of how future automated systems would function, and they expect to need new models to be able to predict the directions of change in vehicle miles traveled, given the uncertainties about traveler decision making. In the absence of real data, it is hard to make reasonable predictions and to educate the decision makers, especially at the elected official level.

Energy and Demand

Existing modeling tools and traveler behavior models are not well suited for representing a future with automated road vehicles, and new modeling paradigms are likely to be needed. Estimates of future road traffic energy consumption with road vehicle automation are highly uncertain because of these limitations. This group was also concerned about the potential for automation to produce highly wasteful applications, such as driverless billboards roaming the streets with advertising displays.

8 General Cross-cutting Observations

As the field of road vehicle automation has advanced and the level of knowledge of the issues has grown over the past several years, the areas of emphasis within the Automated Vehicles Symposium have shifted. In this most recent meeting, several general observations are worth noting:

Many of the key challenges were brought up in a variety of breakout discussions, so these are truly important cross-cutting issues that need serious attention throughout the community interested in road vehicle automation:

- Understanding driver behavior was one of the most frequently mentioned topics, so the in-depth human factors research on this topic, especially associated with how to make safe transitions from automation to manual control, will remain an important need.
- Understanding travelers' trip-making behavior became a more prominent topic than in the past. It is important to understand the propensity of travelers to choose an automated system and also to understand how the use of automation may change decisions about how frequently and how far people will be willing to travel for various purposes.
- Public agency attitudes and actions were an important topic for many breakouts because of the complicated public-private interactions that are almost inevitable in the road transportation domain. The public policies and investments (or shortfalls in investment) are likely to have a strong influence on the adoption of new automated systems and on their effectiveness in improving transportation system performance.
- The growth in the market for automated vehicle systems was one of the most frequently cited uncertainties, since the existing projections have been anything but consistent. Virtually all measures of effectiveness will vary widely depending upon whether automation systems enter the vehicle market gradually, like most previous vehicle technology changes, or whether they produce abrupt and revolutionary changes to consumers' decisions about vehicle purchase or usage.
- Another area of large uncertainty is the anticipated safety and performance of the automated vehicle technologies, and how rapidly those will improve. The world will look very different if those advance gradually from today's vehicle systems or if they make a revolutionary leap forward.
- It would have been easier to conduct the breakout discussions if there had been a well-defined and agreed upon catalog of automated vehicle operational concepts and terms so that everybody would understand each other clearly when discussing their ideas. Unfortunately, these do not exist yet, so the diverse concepts and terminology sometimes become barriers to mutual understanding.
- Multiple breakout groups noted the challenge of communicating clearly with the general public, planners and elected officials about road vehicle automation systems because of the technical complexity of the issues, the lack of standard

concepts and terminology and the widespread and misleading media hype surrounding the topic.

One important topic that was largely overlooked in the previous years' meetings but became much more prominent this year was the cyber-security challenges for automated vehicles. Although this issue transcends automation and is already a concern for all modern vehicles and for the new generation of connected vehicles, it will be seen to be even more important for automated vehicles among the general public because a computer will be doing the driving rather than a normal human driver.

Another broad concern that applied across virtually all topics was the need for credible research, models and demonstrations to reduce the large uncertainties that currently surround automation. People appear to be craving authoritative information about the technical, human factors and policy issues so that they will be better able to visualize and assess how the future will work with more widespread use of automation in road vehicles.

Part I
Public Sector Activities

A National Project in Japan: Innovation of Automated Driving for Universal Services

Hajime Amano and Takahiko Uchimura

Abstract In 2014, the Japanese government initiated a research and development project on automated driving systems. Background, scope, focuses and expected outcome of the project are described in this paper. Deployment of the Intelligent Transport Systems has been actively promoted by the Japanese government in the past 20 years. Technological and operational platforms essential for automated driving systems were formulated as a result of collaboration among public agencies, industries and academia. Application of automated driving technologies is expected to contribute to overcome societal challenges, such as aging society, in addition to road traffic safety, efficiency and enhanced mobility.

Keywords Automated driving · Dynamic map · Cooperative system · GNSS · Human factors · Active aging

1 Outline of the Project

The Japanese government has set out strategies to revitalize Japanese economy and science, technology and innovation. These are closely linked to each other. Under those strategies, a new R&D program was created named Cross-Ministerial Innovation Promotion Program (SIP). Ten projects started in 2014 under SIP. One of them is a project on automated driving systems lead by Dr. Hiroyuki Watanabe as the Program Director [1].

H. Amano (✉) · T. Uchimura (✉)
ITS Japan, 2-6-8 Shibakouen, Minatoku, Tokyo, 105-0011 Japan
e-mail: h-amano@its-jp.org

T. Uchimura
e-mail: t-uchimura@its-jp.org

G. Meyer and S. Beiker (eds.), *Road Vehicle Automation 3*,
Lecture Notes in Mobility, DOI 10.1007/978-3-319-40503-2_2

1.1 Objectives

The objective of the research and development program named SIP was to revitalize Japanese economy and put the country back on the track of sustainable growth by gaining competitive advantages in science and technology. The objective specific to the automated driving project is to provide road traffic safety, environmental sustainability and universal transportation services with special attention to one of the most serious challenges for Japanese society, aging and declining population.

We named the automated driving project as SIP-adus, which stands for innovation of Automated Driving for Universal Services. Inclusive society, where diverse people in diverse communities actively participate in generating values, will enhance both wellness of individuals and economic development. Automated driving technologies integrated with social innovations should provide everyone with mobility to fully exercise his or her capacity, enabling sustainable development of the society [2].

1.2 Scope

We are looking at the evolution of vehicles in the following way. Built-in features of driving assistance are already in the market and getting popular. Cooperative systems have been in nation-wide operation for some years in Japan. Those are integrated into highly automated systems and moving forward to fully automated driving.

Piecewise implementation of technologies, infrastructure and organizational structure will lead us to feasible and sustainable automated driving systems (see Fig. 1).

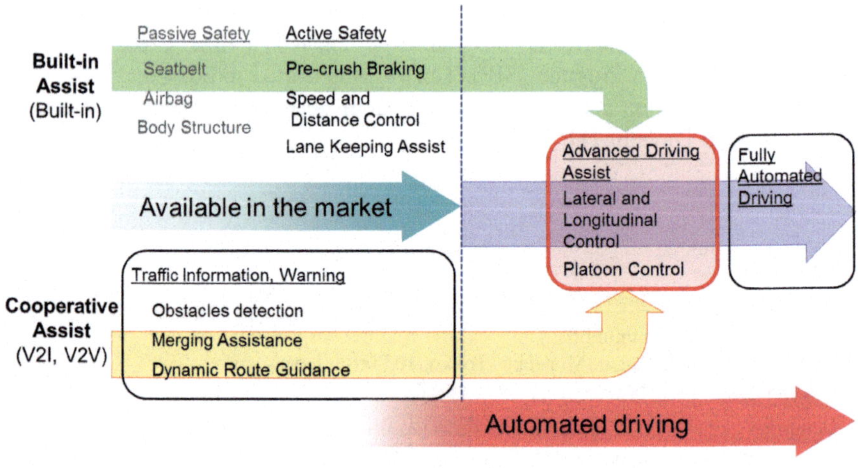

Fig. 1 Evolution of vehicles: connected and automated systems

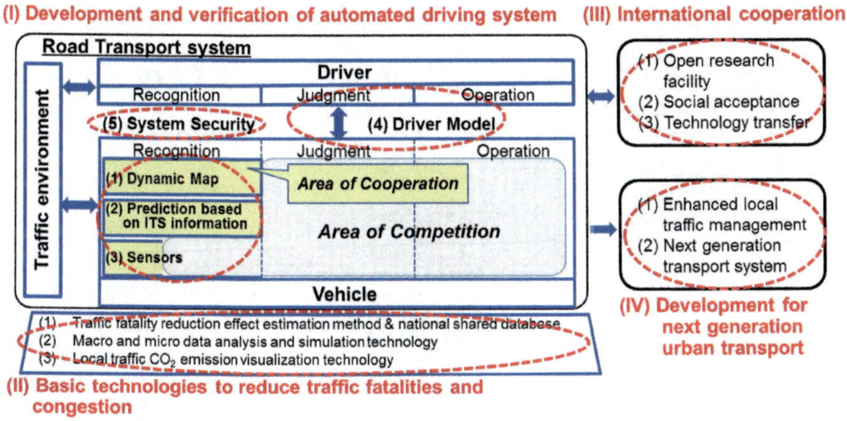

Fig. 2 Scope of SIP-adus project

Automated driving will be realized integrating on-board technologies, precise digital map, data acquisition through radio communication and global positioning. On-board technologies are already in product level competition. Auto manufacturers are demonstrating their technologies and announcing near future products. Therefore, the scope of SIP-adus does not include on-board technologies nor development of prototype automated cars. We are focusing on areas of cooperation, i.e. dynamic map, connected vehicles, human factors, impact assessment, next generation transport, security and international cooperation (see Fig. 2).

2 ITS Deployment as Technological and Operational Platform

The automated driving system project is designed on the portfolio of already operating Intelligent Transport Systems. Deployment of those systems was initiated in 1996 in 9 areas (see Fig. 3). Examples of those systems are described in this chapter from an organizational point of view, which will become foundation of automated driving systems.

2.1 *Electronic Toll Collection*

Operation of Electronic Toll Collection started in 2001 in Japan and soon expanded to nation-wide operation. Among a variety of requirements to guarantee the level of reliable operation acceptable for customers, interoperability between roadside equipment supplied by multiple manufacturers and on-board equipment also

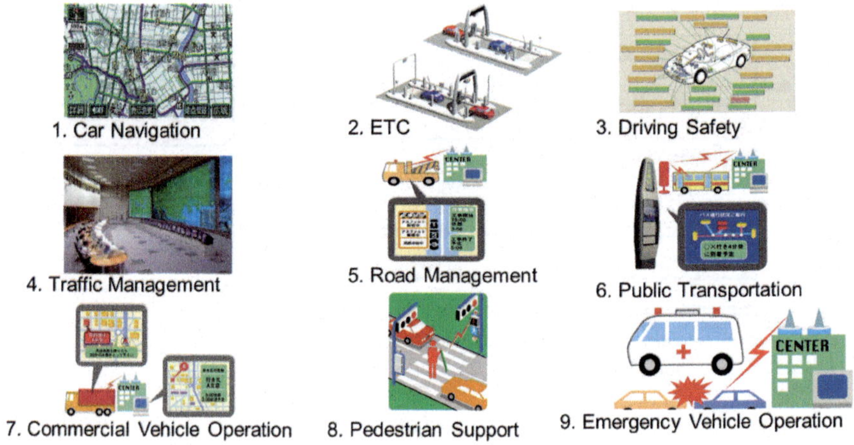

Fig. 3 ITS deployment in 9 areas

supplied by as many manufacturers, and security of every step of transactions were the most challenging ones. For interoperability, standardization of technical specifications was not sufficient to guarantee interoperability. Manufacturers cooperating with the road operators shared information about potential failures and established methodologies for testing conformity to the radio regulation and interoperability. An organization specialized in radio equipment testing has expanded its scope to cover all the related tasks. For security, latest data encryption and authentication technologies were employed. However, dynamic security key management is very new for civilian services. After long and intensive work under collaboration of public and private sectors, a new organization was established for security management of ITS applications. Now, the same framework is applied to vehicle to infrastructure cooperation services, which is one of the important reasons why nation wide operation of cooperative services started so soon in Japan.

2.2 *Car Navigation*

In late 1980s, auto manufacturers and on-board equipment suppliers started to put their first generation of car navigation systems on market. They were already aware that individually developing a digital map database was not a realistic solution. They chose to cooperate with the competitors and the government. A government agency, Geospatial Authority of Japan, has a database of three-dimensional survey of the country. Japan Digital Road Map Association was established in 1988 to develop and maintain a database of digital map and topological structure of road network with unique location reference IDs for a variety of ITS services. Both public and private sectors share the database as common basis and financially

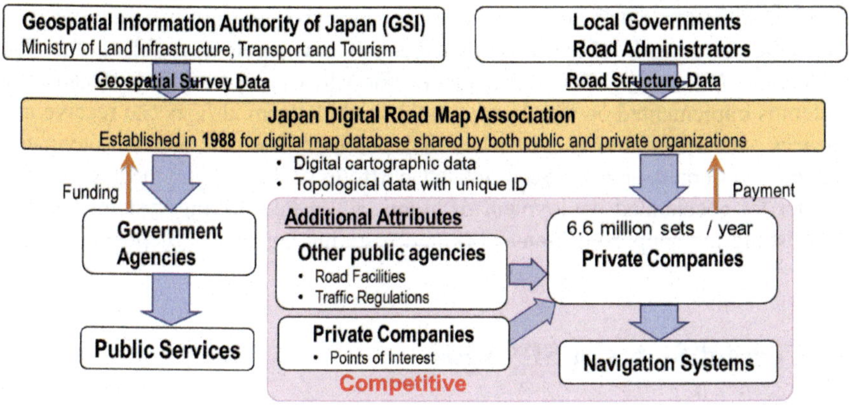

Fig. 4 Digital map for car navigation

support the activity. Private companies are competing by integrating additional attributes for car navigation services. This scheme (see Fig. 4) maintains the balance between cooperation for common basic database to be shared and competition for private service providers to develop creative services at lower cost for their customers.

2.3 *Real-Time Traffic Information Service*

Real-time traffic information service, named Vehicle Information and Communication System (VICS), started in Japan in 1996. Traffic information from highway operators and traffic police is integrated at the Traffic Information Center. The data are coded and broadcast by the VICS center. The car navigation system decodes and displays real time traffic information on the navigation screen overlaying the digital map. The data include level of congestion, road closure and alerts from public agencies. Because common digital map and referencing scheme are shared among all the related public agencies and car navigation system manufacturers, real-time traffic information is properly shown on any mobile terminals.

2.4 *Vehicle to Infrastructure Cooperation*

Vehicle to Infrastructure cooperation services have been in operation since 2011. There are 1,600 radio beacons on expressways and a similar number on arterial roads. Organizations created for earlier ITS deployment are integral parts of the cooperative systems. Since the same radio spectrum and communication protocol as those for Electronic Toll Collection are used, interoperability has already been assured. Although a new set of technologies for cyber security was implemented, the same framework of security management was applied as Electronic Toll Collection. A location referencing scheme for VICS is also utilized. Because the system is implemented on the already established platform, drivers can receive new services with minimal extra cost when they purchase new cars with car navigation and ETC on-board device. Accumulated portfolio of ITS deployment will be an enabler for accelerated deployment of highly automated driving systems for both establishing the supporting framework and fostering customer acceptance.

3 Current Status of SIP-Adus

Some of the progress SIP-adus made at an early stage of the project is described in this chapter. In other areas, project tasks are still in design stage.

3.1 Modeling Road Environment

We are searching for the structure of dynamic map to be built on a shared map database. We are discussing layers with different time frame; static, semi-static, semi-dynamic and dynamic. Probe data collected by the vehicle sensors will be integrated into the database to update the dynamic map [3]. A prototype of basic map layer has been developed for the target area of early deployment, Tokyo bay area (see Fig. 5). The data are shared among the project members and being evaluated. We are going to build prototype of upper layers with semi-dynamic data as the next step.

3.2 Locating Vehicle Position

An automated vehicle will decide, which way to go, matching dynamic map, on-board sensor outputs and global positioning system readings. However, a sufficient number of GPS satellites are not always visible and the accuracy is not as good as we expect. So, we are evaluating other technology options, combination with other sets of GNSS satellites and accuracy enhancement using additional signals (see Fig. 6).

According to our measurement, we obtain lower accuracy with GPS alone. If we combine the data with those from another set of satellites, such as the Quasi-Zenith Satellite System, and enhancement, we obtain better result. We will continue searching for right balance of accuracy requirements for dynamic map, on-board sensors and global positioning system.

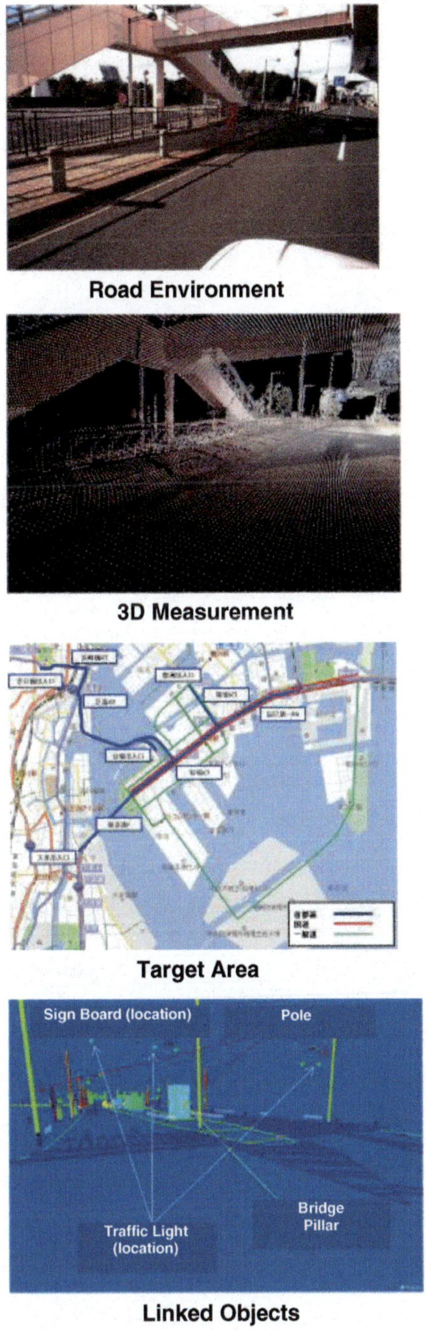

Fig. 5 Prototype of dynamic map

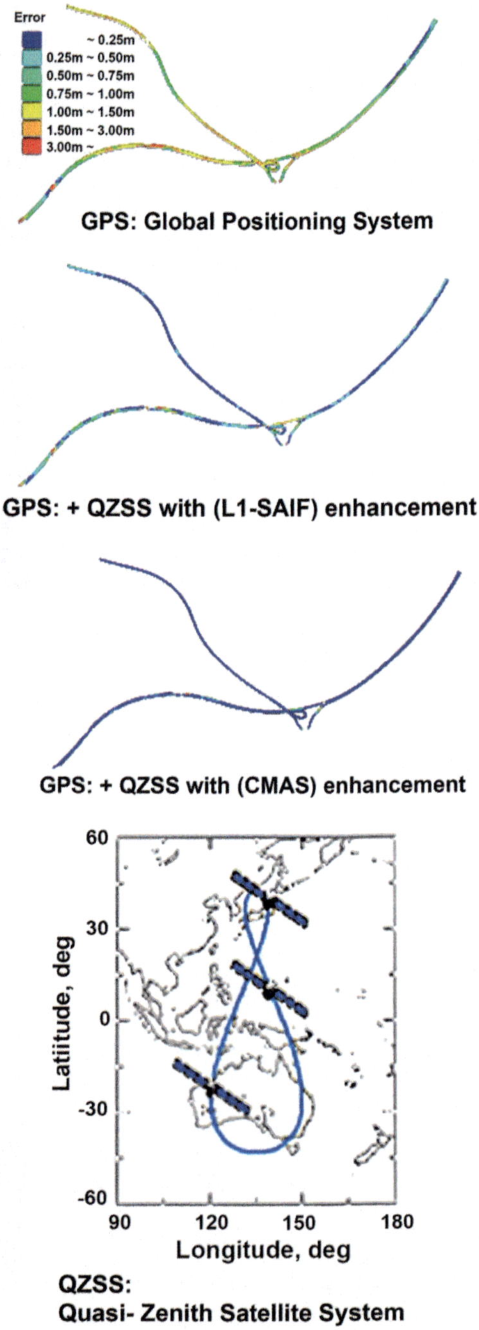

Fig. 6 GNSS accuracy

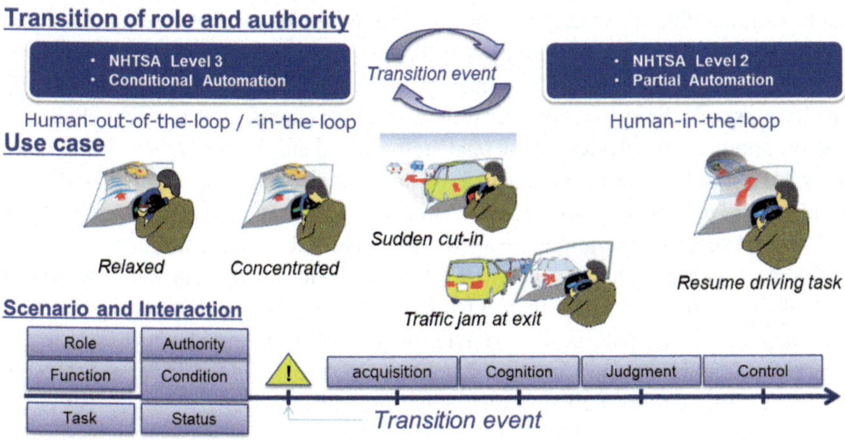

Fig. 7 Analytic approach to human factors

3.3 *Human Factors*

Human factors are also an important area of cooperation, because the implementation of automated driving technologies in commercial products is expected to happen stepwise. The operation of the vehicle will be shared between vehicle control system and human driver until fully automated driving becomes available. There are transitions of roles between vehicle control system and human driver (see Fig. 7). We identified important cases and we are analyzing a series of events, which trigger transition, along the timeline. We are going to observe human behaviors using Driving Simulators and try to find a set of rules acceptable for both human drivers and system design points of view [4].

4 Consideration on Benefits for Japan

Automated driving systems will contribute to solve problems directly related to road traffic, such as safety and efficiency. However, we will have benefits from automated driving technologies in more fundamental societal challenges, such as aging society.

4.1 *Road Safety*

Road traffic fatality was more than 16,000 per year around 1970 in Japan. Better road facilities, education and enforcement were effective countermeasures on those days. But the fatality started to increase again in 1980s for scoring economic activities. This time, new vehicle technologies and Intelligent Transport Systems significantly contributed to reduce fatality. Although the total number of road traffic fatality keeps declining slowly, rapid demographic change poses a new challenge. Today, about 54 % of victims of fatal traffic accidents are 65 years old or older. They are not only victims of the accidents but they also cause accidents. Fatal traffic accidents are classified by the types of violation. The most conspicuous observation is that inappropriate operations by aged drivers caused so many accidents. National Police Agency conducted a survey of families of victims of fatal traffic accidents. One of the highest priorities they wish is better countermeasures for aged drivers and they also anticipate wider penetration of collision avoidance technologies. Therefore, immediate application of automated technologies for driver assistance to the models already on market is our most imminent mission.

4.2 CO_2 Emission Reduction

The shared platform to be created for automated driving, such as dynamic map database, will be utilized by a variety of applications. With traffic information from fixed sensors and moving vehicles combined and detailed digital road map, we can precisely reproduce movement of each vehicle on the computer. Then, we can get the total CO_2 emission volume in the area. Collaborating with traffic engineers from

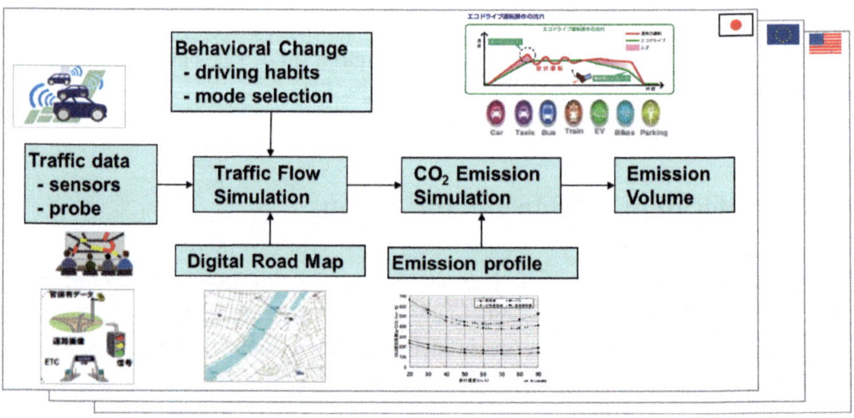

Fig. 8 Quantitative estimation of CO_2 emission

Europe and the United States, we have internationally recognized methodology (see Fig. 8) [5]. As a part of SIP-adus project, we are going to evaluate impacts of automated vehicles on energy consumption.

4.3 Active Aging

Aging and declining population is one of the most serious challenges for Japan. More than 25 % of Japanese population is 65 years old or older now. This share is expected to increase up to 40 % by the year 2050. The current social welfare system no longer works, which assumes that younger generations support all the people of 65 years old and above. We have to mobilize people of this age group to sustain the society. United Nations recognized it and proposed a policy framework: Active Aging. Automated driving technologies are expected to contribute giving people adequate mobility such that they can continuously play active roles in the society. For mega cities, such as Tokyo, where comprehensive coverage of subway network exists, a flexible transit system on the ground level is anticipated. In rural areas, depending heavily on private cars, we need safe driving assist and new compact cars for the older age group to travel [6]. Figure 9 is a conceptual image of Advanced Rapid Transit, to be deployed by the Tokyo Olympic and Paralympic Games in 2020 [7].

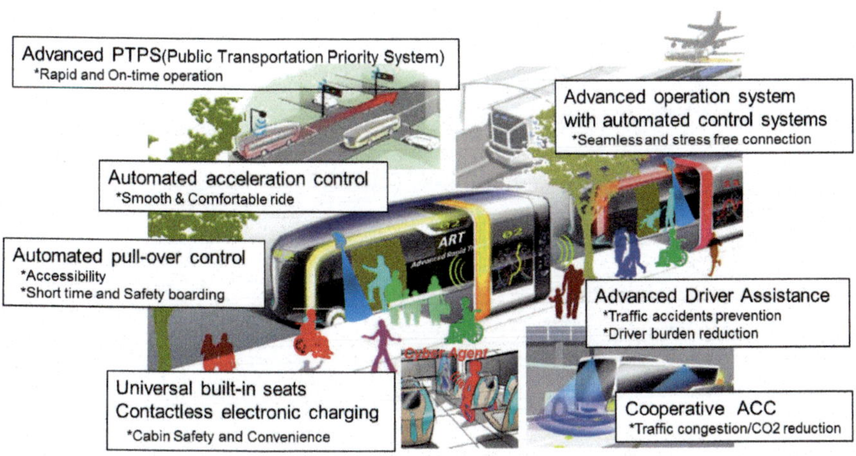

Fig. 9 Advanced rapid transit system

5 Conclusion

Automated driving technologies are actively developed and regulatory aspects are discussed to support automated vehicles to operate. Collaboration to establish a shared framework for interoperability, security management, dynamic map and human factors is also important. We will gain from automated driving technologies not only road traffic but also societal benefits.

References

1. Tanuma T (2014) The National Program for Innovation - Cross-ministerial Strategic Innovation Promotion Program (SIP) - SIS72. ITS World Congress Detroit
2. Amano H (2015) Safe and efficient mobility for active aging society, Session 564, TRB Annual Meeting
3. Shirato R (2014) Introducing the activity for dynamic map in SIP-adus. SIP-adus Workshop, 2014
4. Unoura K (2014) Humans and automated driving systems. SIP-adus Workshop, 2014
5. Kuwahara M, Shladover S et al, Guidelines for assessing the effects of ITS on CO_2 emissions-International Joint Report. Institute of Industrial Science The University of Tokyo, 2013
6. Amano H (2015) SIP-adus: a National Project in Japan, SIS31, ITS World Congress Bordeaux, 2015
7. Kawamoto M (2014) A concept of SIP-adus for the next generation transport, SIP-adus Workshop, 2014

Accessible Transportation Technologies Research Initiative (ATTRI)—Advancing Mobility Solutions for All

Mohammed Yousuf, Jeffrey Spencer, Robert Sheehan and Louis Armendariz

Abstract The Accessible Transportation Technologies Research Initiative (ATTRI) leverages emerging innovations to identify, develop, and deploy new transformative technologies, applications, or systems, along with supporting policies and institutional guidance, to address mobility challenges of all travelers, in particular, travelers with disabilities. ATTRI research focuses on the needs of three stakeholder groups: people with disabilities, veterans with disabilities, and older adults. ATTRI will also develop technological solutions to lower or remove barriers to transportation according to four functional disabilities within these stakeholder groups: visual, hearing, cognitive, and mobility. The technologies potentially leveraged by ATTRI provide almost ubiquitous access to a wealth of real-time situational data sources, including data specific to transportation, municipalities, points of interest, accessibility, and crowd-sourced information.

Keywords ATTRI · Accessible transportation · Emerging technology · Accessibility data · Vehicle automation · Intelligent transportation systems (ITS) · Connected vehicles (CV) · Vehicle-to-infrastructure (V2I) communication

---

M. Yousuf (✉)
Turner-Fairbank Highway Research Center, Federal Highway Administration,
6300 Georgetown Pike, HRDO-04, McLean, VA 2210, USA
e-mail: Mohammed.Yousuf@dot.gov

J. Spencer
U.S. Department of Transportation, Federal Transit Administration, 1200 New Jersey Avenue SE, Washington, DC 20590, USA
e-mail: Jeffrey.Spencer@dot.gov

R. Sheehan
U.S. Department of Transportation, ITS Joint Program Office, 1200 New Jersey Avenue SE, Washington, DC 20590, USA
e-mail: Robert.Sheehan@dot.gov

L. Armendariz
Booz Allen Hamilton, Dranesville, VA, USA
e-mail: Armendariz_Louis@bah.com

G. Meyer and S. Beiker (eds.), *Road Vehicle Automation 3*,
Lecture Notes in Mobility, DOI 10.1007/978-3-319-40503-2_3

1 Introduction to ATTRI

The development and, to an even greater degree, the acceptance and deployment of advanced transportation technologies is a complex process. Entrepreneurs, researchers, and futurists seem to discuss a new technological revolution almost daily. Hundreds of innovators are working to bring transformations in accessible transportation technology to assist people with disabilities and older adults. The Accessible Transportation Technologies Research Initiative (ATTRI) is a joint U.S. Department of Transportation (USDOT) initiative, co-led by the Federal Highway Administration (FHWA) and Federal Transit Administration (FTA), with support from the Intelligent Transportation Systems (ITS) Joint Program Office (JPO) and other Federal partners. ATTRI conducts research to improve the mobility of travelers with disabilities through the use of ITS and other advanced technologies. ATTRI leads the research, development, and implementation of transformative technologies, solutions, applications, or systems for people of all abilities to effectively plan their personal and independent travel.

ATTRI will enhance the capability of travelers to reliably and safely execute independent travel. ATTRI will identify, develop, and deploy new transformative technologies, applications, or systems, along with supporting policies and institutional guidance, to address mobility challenges of all travelers, particularly those with disabilities. ATTRI supports this effort with its mission to remove barriers to transportation by leveraging advanced technology to enable people to travel more easily, affordably, and effectively, regardless of their individual abilities. All applications developed under the ATTRI program shall utilize principles for all users of the transportation ecosystem, whether developing new applications that can be used by all travelers or leveraging existing solutions and enhancing them to meet user needs. As part of this mission, ATTRI is responsible for communicating its goals, objectives, research, and progress to its stakeholders and the general public.

1.1 Background

The USDOT remains a driving force in the future of accessible and inclusive transportation technologies [1]. Transportation plays a critical role in enhancing access to education, family, work, healthcare, leisure, and social activities (Fig. 1).

Research has shown that enhanced personal mobility for travelers with disabilities has a direct impact on labor force participation and increases economic activity. The USDOT acknowledges that transportation is a key driver for these systems and includes enhancing mobility as a strategic priority within the *ITS Strategic Plan (2015–2019)*. The ITS Program looks to explore methods and management strategies to increase system efficiency and improve individual mobility [2].

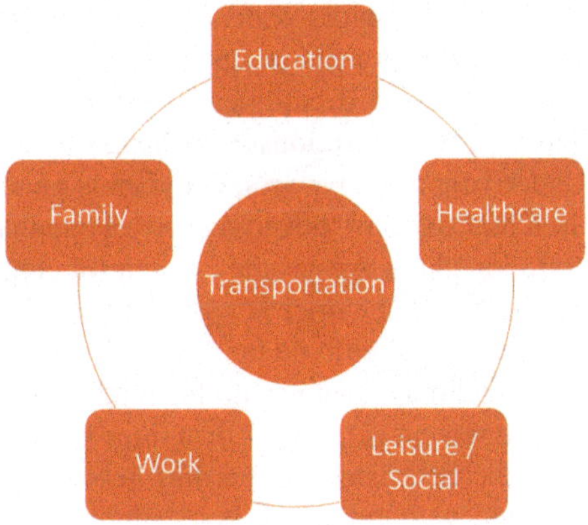

Fig. 1 Role of transportation

1.2 Challenges and Barriers

The barriers facing persons with disabilities, veterans with disabilities, and older adults–whether those challenges are manifested by virtue of vision, mobility, hearing, or cognition–are as diverse as that population itself. Approximately 56.7 million people (18.7 %) of the 303.9 million in the civilian noninstitutionalized U.S. population had a disability in 2010 [3]. Lessons from a survey of persons with disabilities in New Jersey found "76 % of all participants felt strongly that transportation was important for their job search…. When asked whether they left or refused a job offer because of travel difficulties, 25 % mentioned leaving a job and 40 % mentioned refusing a job offer because of travel difficulties" [4]. The global gross domestic product lost annually due to disability is estimated to be between $1.37 and $1.94 trillion [5]. According to a World Health Report, "over a billion people, about 15 % of the world's population, have some form of disability and rates of disability are increasing due to population ageing and increases in chronic health conditions, among other causes" [6]. The population 65 and over has increased from 35.5 million in 2002 to 43.1 million in 2012 (a 21 % increase) and is projected to increase to 79.7 million in 2040. Some type of disability (i.e., difficulty in hearing, vision, cognition, ambulation, self-care, or independent living) was reported by 36 % of people age 65 and over in 2012 [7].

Many barriers often faced by these communities include lack of accessible service; lack of available transportation; lack of signage, maps, and landmarks; navigation difficulties; inconsistent accessible pathways and infrastructure; no accessible amenities (restrooms, benches, etc.); unreliable modes of transportation; and the weather. However, by and large, the barriers can be categorized into three different categories: (1) the actual mechanics of executing an individual's transportation needs, and (2) determining how an individual could execute his/her transportation needs, and (3) communicating transportation innovations to stakeholders who may have vision, hearing, mobility and intellectual impairment.

The first category depends on the nature of the disability(ies) that an individual has. A flight of stairs or steep terrain can inhibit the mobility of those persons with mobility-related disabilities, whereas a four-stage signalized crossing at a busy intersection can seriously challenge those with any and all types of disabilities. Underlying those barriers, our team is mindful that these populations often also face the challenges of poverty. Too often, that poverty is exacerbated by the transportation challenges as they search for employment opportunities to which they can travel. Another barrier is cost. The enabling technologies must be affordable. The technologies should be innovative but simple for the older population to be able to use it without a steep or no learning curve as many of them are technology averse.

The second category is then informed and often defined by the array of challenges from the first category that lies between the traveler's origin and destination. For example, wayfinding can be as simple as helping someone with mobility challenges find a path that most quickly routes them to an elevator at their subway stop rather than first to the stairs. However, wayfinding can be as complex as providing turn-by-turn auditory and sensory instructions through a major international airport for a passenger with a visual disability. There, the wayfinding cannot only be to the passenger's departing gate, but also to an accessible restroom facility—not to mention a bite to eat or cup of coffee. Ultimately, the wayfinding should serve to make the traveler's trip as stress-free as possible.

The third category speaks to the traveler's ability to receive the information. For example, is the text too small for an elderly or vision-impaired person? Can the sound be played at a desired speed as visually impaired individuals can often process auditory messages faster than those not visually impaired? For a person with intellectual impairment, are texts or visuals in plain language easy to understand? Do visuals tell compelling stories that are simple yet informative? Communications to these stakeholders must be very clear in order to be understood and processed.

1.3 ATTRI Vision

The ATTRI program's vision is to enhance the mobility of travelers with disabilities by providing the capability to reliably, safely, and independently plan and execute their travel. The initiative identifies, coordinates, develops, and implements new integrated solutions in advancing such capabilities.

The driving purpose of this initiative is to leverage technological advances to transform the mobility of travelers with disabilities. ATTRI addresses the mobility challenges of travelers with disabilities by identifying, developing, and deploying new applications or systems that support policies and institutional guidance. The initiative focuses on the needs of three stakeholder groups: people with disabilities, veterans with disabilities, and older adults (Fig. 2).

Fig. 2 Targeted populations of the ATTRI program plan

Through ATTRI, technological solutions will be developed to lower or remove barriers to transportation according to four functional disabilities: visual, hearing, cognitive and mobility.

1.4 User Needs Identified

In its ongoing effort to gain valuable information from the public, the USDOT conducted outreach and stakeholder engagement activities, including a series of webinars to identify user needs and challenges faced by a variety of ATTRI stakeholders, including, but not limited to, the three main stakeholder groups that were previously listed. The webinars were interactive and walked participants through several scenarios to gather responses and feedback. These webinars were very interactive and participants shared a wealth of information to assist us in identifying and growing our understanding of user needs.

Based on the barriers identified earlier, a set of needs can be developed. These include the need for amenity information ahead of time (availability of accessible restrooms, benches, shelter, etc.), real-time transportation information (to help with connections), (traveler helpline or customer service), consistent accessible pathways, destination information (entrances, elevators, layout, etc.), mapping (for directions), roadway/pathway real-time conditions (during adverse weather) and personal care attendant or other assistive services. Using innovative and simple technology can help address almost all needs, but there are issues with technology as well that need to be kept in mind, while developing applications.

For example, applications need to work with existing systems in place, have a systematic maintenance schedule, be reliable, and most importantly be immunized to data confidentiality breaches/theft. Technology initiatives can address several problems these target populations face on a daily basis. For example, a network of ITS technologies can be used to reduce or eliminate the safety risks associated with an at-grade street crossing by communicating to vehicles the presence and needs of a crossing pedestrian. Automated vehicles and personal mobility can serve to elegantly tunnel through the entire landscape of barriers by providing technologies whose demands of the traveler are independent of the traveler's abilities—namely, the traveler may only need to know where they want to go and need to provide no other input, physical, cognitive or otherwise. Robotics and artificial intelligence can be invaluable resources for wayfinding as assistants to the traveler. Accessible data can be the foundation of all wayfinding applications as those data codify the existence of the barriers of individual traveler's objective function, enabling personalized and mathematically simple path optimization. These technologies should serve to reduce how onerous various features are along a differently-abled traveler's path and also reduce how onerous the path is on the whole provided each individual's unique set of abilities and challenges.

1.5 Collaboration

As a USDOT joint research and development initiative, ATTRI is co-led by Federal Highway Administration (FHWA) and Federal Transit Administration (FTA), with support from ITS Joint Program Office (ITS JPO) and other Federal agencies, such as the National Institute on Disability, Independent Living, and Rehabilitation Research (NIDILRR). The USDOT recognizes the interdisciplinary nature of accessible transportation research and actively seeks opportunities to leverage resources, accomplishments, and knowledge transfer, both within USDOT as well as across Federal agencies. ATTRI has established collaborations with key federal partners and other organizations to leverage technologies and innovations from Federal ITS and related disability research and development activities. Key partners include the Interagency Committee on Disability Research (ICDR); the U.S. Access Board; the U.S. Army Tank Automotive Research, Development, and Engineering Center (TARDEC); as well as other public and private organizations.

2 ATTRI Technology Research Solutions

Potential applications to be developed under ATTRI will leverage advances in vehicle and infrastructure-based technologies, automation, robotics, and wireless communication. Other USDOT research program areas and emerging innovations (such as assistive robots and crowdsourcing) could help produce seamless transportation capabilities. These program areas include vehicle automation; vehicle-to-vehicle (V2V), vehicle-to-infrastructure (V2I), and vehicle-to-pedestrian (V2X) communications; the Veterans Transportation Community Living Initiative (VTCLI); and Mobility Services for All Americans (MSAA). Five technology areas have emerged as potential ATTRI focus areas to improve transportation for people disabilities: wayfinding and navigation, assistive technologies, automation and robotics, data integration, and enhanced human services transportation (Fig. 3).

Fig. 3 ATTRI technology research solutions

2.1 *Wayfinding and Navigation Solutions*

This area consists of exploration and development of situational awareness and assistive navigation solutions that can provide obstacle avoidance and intelligent wayfinding capabilities in indoor and outdoor environments. These solutions assist with waypoint navigation, path planning, and advanced warning of events by using Global Positioning System (GPS), geographic information system, and ITS equipment and technologies. These applications can recognize and detect stationary objects (e.g., doors, elevators, stairs, crosswalks, and traffic lights); read and recognize important text and signage based on a user's query; and detect, track, and represent moving objects and dynamic changes to a traveler's environment (e.g., people, shopping carts, doors opening, and moving vehicles). Wearable sensors, such as cameras, three-dimensional orientation devices, and pedometers, may be used in conjunction with a display unit to provide auditory and tactile guidance.

2.2 *ITS and Assistive Technologies*

The broad range of wireless and sensor-based communications and information technology employed in ITS, combined with a number of other assistive technologies, can create new innovative accessible transportation solutions. This will include the traditional accessible, assistive, and adaptive devices that currently help with daily living activities and new nomadic or personal devices. Together, these technologies will help track the user's movements, infer map information, and discover key sensor signatures to create routes and provide information such as data for caregivers and virtual attendants in different accessible communication formats: audible, tactile, and haptic. The devices used may include new innovations from the

"Internet of Things" (IoT) that are applicable to wearable technology, such as wrist bands, glasses, or clothing. These technologies will also integrate with vehicles, infrastructure, and pedestrians using dedicated short range communication (DSRC) or other communication technologies to provide V2V, V2I, and V2X communications. This will allow for connectivity throughout a trip. This area will also explore other emerging technologies within the connected vehicles, connected automation, and connected cities initiatives under the USDOT's connected vehicle research program.

2.3 Automation and Robotics

Automated vehicles and robotics are expected to improve mobility for those unable or unwilling to drive, and enhance independent and spontaneous travel capabilities for travelers with disabilities. One area of particular interest is exploring the use of vehicle automation to solve first mile/last mile mobility issues, and possibly providing connections for all travelers to existing public transportation or other transportation hubs. Applications in this area may also include collaborative robots that not only assist with daily life activities such as walking, but also work with individual travelers and human transportation services to provide related concierge services at different stages of their travel, and hence improve personal mobility across the transportation network. Machine vision, artificial intelligence, assistive robots (potentially partially humanized), and facial recognition software are included to solve a variety of travel-related issues for people with disabilities in vehicles, devices, and terminals. These technologies can create virtual caregivers/concierge services and other such applications to guide travelers and assist decision making.

2.4 Data Integration

This technology area includes solutions that enable the integration of data and information systems to create new accessible transportation applications. This technology area has two main aspects: information that travelers with disabilities need, and information that travelers with disabilities can provide. Travelers with disabilities need in-depth accessibility information about points of interest, infrastructure, facility amenities, and potential obstacles, integrated with maps and other information for their intended route. In addition, a traveler can provide his or her specific information to build a standardized user profile with accessibility needs that allows for location-based services, both locally and nationally. Based on the user profile, applications can be developed to alert relevant authorities in advance of a user's trip requiring special accommodations, such as a wheelchair at the airport.

2.5 *Enhanced Human Services Transportation*

The focus of this technology area is real-time, multimodal trip and services planning and traveler decision support applications that assist travelers with finding and choosing accessible transportation solutions that best meet their mobility needs. This may include pre-trip information and planning tools that integrate multi-modal options into a complete origin-to-destination trip. Applications in this area could include an integrated payment system, where travelers can use the same smart card or mobile app to pay for various types of transportation, mobility options, and parking. Other applications of interest include linking paratransit, demand-response transportation, and fixed-route transit in order to increase flexibility and options for travelers with disabilities.

3 ATTRI Three Phase Program Plan

ATTRI is being implemented in three phases. The phases include: Exploratory and User Needs Research; Innovation, Prototype Development and Testing; and the Demonstration Phase.

Each phase will prioritize particular focus areas. These areas are as follows (Fig. 4).

Phase 1: Exploratory and User Needs Research:

- Collaboration building
- Strategic plan and roadmap
- Stakeholder engagements and user needs assessment
- State of the practice / innovation scan

Phase 2: Innovation and Prototyping:

- Accessible transportation applications selection
- Institutional and policy issues assessment
- International research coordination
- Social-economic impact

Phase 3: Demonstration:

- ATTRI joint demonstrations
- Deployment guidance
- Evaluations

Fig. 4 ATTRI three phase program plan

4 ATTRI Foundational Considerations

A number of entrepreneurs, researchers, and futurists are working to bring transformations in accessible transportation technology to assist people with disabilities and the travelers in general. The acceptance and impact of such applications will be measured by the ease of use and the simplicity, more importantly, these applications need to communicate and act in different ways to different users yet accomplish the same task of getting someone from point A to point B in different cities, towns and neighborhoods. Similar or repeatable traveler experience, irrespective of the mode of transportation or destination, can be achieved by designing interfaces and output mechanisms that interact the same way across different settings with specific foundational considerations in mind. Through outreach and stakeholder engagement activities of ATTRI, including several public workshops, a User Needs Assessment, a Technology and Innovation Scan, four (4) foundational considerations have emerged as necessary elements of accessible transportation applications to achieve a unified travel experience that meet the diverse needs of travelers with mobility, vision, hearing and cognitive disabilities. These foundational considerations are included, in the development of four (4) identified ATTRI priority application areas: (1) Smart Wayfinding and Navigation systems; (2) Pre-Trip Concierge and Virtualization; (3) Shared Use, Automated Vehicles; and (4) Safe Intersection Crossings. More specifically, any application developed under ATTRI will have considered and incorporated the foundational considerations described below.

4.1 Standard Accessible Data Platform

Data sharing and standardization is critical in developing applications which aspire to enhance the personal mobility of those with the greatest needs. Service platforms and providers, using interoperable communications, can push and pull real-time data to communicate in an efficient, succinct, adaptable, and understandable manner to meet individual user needs. Technology applications to be considered for development under the ATTRI program would like to make use of almost ubiquitous access to a wealth of real-time, situational data sources, including data specific to transportation systems, municipalities, points of interest, crowd-sourced information, and accessibility data available from different sources, including individual profiles.

4.2 Universal Design Standards

Universal design standards incorporate a philosophy that promotes the applicability of a technical solution to the needs of all user groups. An accessible Information

Communications Technology (ICT) product or service is one which can be used by all its intended users, taking into account their differing capabilities. A person's ability to use technology may be impaired due to various physical, sensory, emotional or cognitive disabilities [8]. All applications developed under the ATTRI program shall utilize universal design principles, whether developing new applications that can be used by all travelers or leveraging existing solutions and enhancing them to meet the needs of all users.

4.3 *Integrated Mobile Payment*

Integrated mobile payment systems typically incorporate interoperable electronic fare payment that can be utilized across various modes of transportation by all travelers including those with disabilities, at all times, and for multiple consumer purposes. The vision for a multimodal integrated payment system is to deliver, for travelers in the transportation ecosystem, the ease of use and convenience that comes from an electronic payment system and extend that ease across modes and through institutional and technical collaborations. An integrated fare payment is especially important for travelers with disabilities, particularly those with visual disabilities where they may have difficulty in distinguishing between different payment types or those with cognitive disabilities that may become confused by multiple payment requirements.

4.4 *Leverage Existing Technologies*

To maximize the impacts of limited ATTRI resources and respond most effectively to the needs of all users and stakeholders, any application being developed under ATTRI should leverage existing promising technologies, including but not limited to Intelligent Transportation Systems (ITS), on-demand technologies, data standards, innovative smartphone and mobile technology, and transportation and other assistive and enabling technologies, operations, and/or techniques whether currently being pursued in research, or readily available in the market. For example, can wearable technologies such as smart watches integrate accessibility needs within health monitoring applications, or provide information to remote caretakers? Can emerging wheelchair technologies such as Wii or Ogo be used in conjunction with nomadic devices to increase personal mobility vehicle options for first mile last mile mobility? In addition, partnerships with other current research or deployment efforts can help conserve resources and avoid duplicative research.

5 Moving Forward

As the second phase of the initiative begins, ATTRI is addressing a significant transportation problem in a comprehensive way. With increasing stakeholder engagements, workshops, and Request for Proposal (RFP) procurements, the ATTRI program is positioning itself to capitalize on potential large-scale opportunities. ATTRI research will improve the mobility of travelers with disabilities and provide enhanced capabilities for all travelers to reliably and safely execute independent travel. Through ATTRI research and associated new technology solutions, users in the transportation ecosystem will realize safe travel, using real-time situational awareness and information sharing [9].

References

1. Eric L (2015) The US DOT leaders in accessible transportation technology?. Huffington Post. 17 July 2015
2. Barbaresso J, Cordahi G, Garcia D, Hill C, Jendzejec A, Wright K (2014) USDOT's Intelligent Transportation Systems (ITS) ITS Strategic Plan 2015–2019. Intelligent Transportation Systems Joint Program Office, Washington, D.C
3. Brault M (2012) Americans with Disabilities: 2010, U.S. Department of Commerce, U.S. Census Bureau, Current Population Reports, pp 70–131. July 2012
4. Lubin A, Deka D (2011) The role of public transportation as a job access mode: lessons from a survey of persons with disabilities in New Jersey, TRB Paper, vol. 12–1480. 28 July 2011
5. Metts R Disability Issues (2000) Trends and Recommendations for the World Bank, 1.2.2 Estimates of GDP Lost Due to Disability. February 2000
6. World Health Organization (2015) Disability and health webpage. www.who.int/mediacentre/factsheets/fs352/en/
7. Administration for Community Living (2014) Administration on Aging, U.S. Department of Health and Human Services, A Profile of Older Americans: 2014, Washington, D.C
8. Regional ICT Champions (2008–2011) What does ICT accessibility mean? NAVCA. www.ictchampions.org.uk/2009/05/what-does-accessibility-mean/
9. Material produced in this report can be found online at the U.S. Department of Transportation, Intermodal Research: Accessible Transportation Technologies Research Initiative webpage. www.its.dot.gov/attri/

DOE SMART Mobility: Systems and Modeling for Accelerated Research in Transportation

Reuben Sarkar and Jacob Ward

Abstract New paradigms emerging in transportation and information and communication technology create opportunities to better understand and leverage the interactions between travelers, vehicles, and the built environment to reduce greenhouse gas emissions and save energy. The U.S. Department of Energy's SMART (Systems and Modeling for Accelerated Research in Transportation) Mobility Initiative recognizes and harnesses these megatrends by elevating DOE's traditional transportation energy focus beyond the vehicle component technology level to transportation-as-a-system analysis, modeling and simulation, and applied research and development in 5 interrelated topics: connected and automated vehicles, mobility decision science, urban science, vehicles and infrastructure, and multi-modal.

Keywords Connectivity · Automation · Transportation · Vehicles · Mobility · Energy · Emissions · Decision science · Urban science · Multi-modal · Infrastructure

1 Initiative Description and Impact

New paradigms emerging in transportation and information and communication technology create opportunities to better understand and leverage the interactions between travelers, vehicles, and the built environment to reduce greenhouse gas emissions and save energy. Research led by key stakeholders in the US and abroad suggests vast energy implications related to transportation system efficiencies. Vehicle connectivity and automation alone has been projected to have a potential impact between −90 % and +200 % to greenhouse gas emissions relative to the U.

R. Sarkar · J. Ward (✉)
U.S. Department of Energy, 1000 Independence Ave., S.W.,
Washington, D.C. 20585, USA
e-mail: jacob.ward@ee.doe.gov

R. Sarkar
e-mail: reuben.sarkar@ee.doe.gov

G. Meyer and S. Beiker (eds.), *Road Vehicle Automation 3*,
Lecture Notes in Mobility, DOI 10.1007/978-3-319-40503-2_4

S. Energy Information Administration's business as usual scenario [1] by 2050 according to a preliminary DOE national laboratory study. This opportunity is underscored by increasing urbanization. Cities are responsible for 75 % of global CO_2 emissions and their populations are projected to increase from 3.5 billion people today to 5 billion people in 2030 [2].

R&D to develop lower-carbon fuels and clean, efficient vehicle technologies such as electrified powertrains remains critical to meeting national energy and greenhouse gas objectives. However, this focus on individual vehicle-level efficiency doesn't take into full consideration the efficiencies and greenhouse gas reductions that could be extracted from better utilization of transportation assets (including vehicles) as part of a broader integrated transportation system. An understanding of how transportation assets, travelers, and the transportation system interact and influence each other using multi-scale, multi-system models is required to evaluate and optimize the overall system. Currently there is a gap in R&D programs to directly exploit these opportunities, which are likely to be essential to meet IPCC's estimated reduction needed in greenhouse gas emissions to keep global temperature increase below 2 °C [3]. This challenge and opportunity spans a wide array of topics and potential undertakings that can be categorized into the following five interrelated pillars, which are also shown in Fig. 1:

- Connectivity and Automation
- Mobility Decision Sciences
- Urban Sciences
- Vehicle and Infrastructure
- Multi-Modal Transportation.

Fig. 1 The five pillars of the proposed DOE SMART mobility framework

These five pillars represent opportunities to save energy at the macro, or transportation-system, level and need to be further refined and informed by modeling and analysis. Research to address these pillars of Systems and Modeling for Accelerated Research in Transportation (SMART) Mobility will require the combined effort of federal agencies, national laboratories, universities and key industrial partners and stakeholders. This will support robust analytical and foundational efforts to define and build-up the Transportation-as-a-System constituent parts and frame the R&D priority opportunities. In addition to models, tools, and applied analysis, support for focused technology demonstrations in conjunction with cities or states will be needed to spur commercialization and inform future activities across the transportation sector.

Although represented as one of the five pillars, the onset of fully connected and automated vehicles (CAVs) is a driving force in all pillars that may cause technology, systems, and business models to reach key tipping points, ushering in wide-scale changes in surface transportation markets. CAVs alone, in isolation from the other pillars, provide significant vehicle level opportunities for energy efficiency advancements. When intersected with the other pillars, CAVs become both a disruptive business force as well as an accelerant of potential energy impacts from other perspectives (both positive and negative). For this reason CAV technology is treated not only as a foundational pillar, but also frequently discussed in other pillars as a system level catalyst.

Timeliness for undertaking research and development in this focus area is critical, as many industry players have already invested significant resources towards incorporating connectivity and automation into vehicles of all classes, as can be seen with active lane-departure systems and self-parking vehicles. Traditional vehicle manufacturers are also being joined by new industry players. The transformative impact resulting from convergence of market sectors and social and technology trends points towards a near-term opportunity for proactive and cross-disciplinary engagement in these areas to ensure that energy savings and greenhouse gas reduction potentials can be maximized. Examples of current industry and market activities include:

- Passenger vehicle manufacturers such as GM, Volvo Car Corp., Tesla and many others are incorporating increasing levels of CAV features (such as lane-keeping, lane-changing, adaptive cruise control, auto-parking, etc.) into their product offerings, particularly for high-end vehicle models. At the same time, heavy-duty players such as Freightliner, Volvo Trucks, Daimler Trucks and Peloton Technologies have major projects to develop and test truck-platooning technologies [4]. Beyond traditional vehicle manufacturers, technology companies such as by Google are pushing the envelope on fully-automated driving across the spectrum of road situations and conditions.
- Processor, sensor, and data companies (e.g. Cisco, Qualcomm, Verizon) have made vast investments into the “internet of things”, which is providing increased

data processing capabilities, and much greater connectivity and personalization of mobility, particularly in cities.

- Mass-market 200-mile range battery electric vehicles (e.g. Chevy Bolt, Tesla Model 3) are in development and wireless charging is expected to be introduced on next generation EVs, which suggests the need for a concerted public-sector effort to align increased vehicle capabilities with charging infrastructure technologies and investments. Similarly, the introduction of consumer hydrogen fuel cell electric vehicles (e.g. Toyota Mirai, Hyundai Tucson) and other non-petroleum fuels points to a similar need to optimize infrastructure investment and placement.
- Freight transport is becoming increasingly distributed and same-day delivery by companies such as Amazon and Google are changing overall transport logistics, which will be further affected by automated technologies.
- Dynamic ridesharing companies such as Uber and Lyft are changing transportation dynamics and ownership models with the stated intention to offer a service that is more affordable than vehicle ownership, which suggests a potential future for automated movement of passengers and freight.

This level of activity suggests the need to pursue research that incorporates individual and cross-cutting impacts of the five SMART Mobility pillars in order to provide additional tools and levers to meet national greenhouse gas and energy independence targets. The scope of each of these pillars, priority R&D needs, and completed foundational work is described below in order to provide greater detail on both the impact and opportunity for harnessing transportation system energy efficiencies.

2 Connectivity and Automation

Connected and automated vehicles are potentially one of the most transformational technologies introduced to transportation in decades, presenting new opportunities for both vehicle technologies and transportation business models. While most research to date has focused on safety, the impact on energy consumption and greenhouse gas emissions has not been explored beyond high-level, preliminary analysis. There is much uncertainty regarding technological progress, behavioral changes, demographic changes and mobility policy implementation involving CAVs. Preliminary work conducted by multiple U.S. Department of Energy (DOE) National Laboratories has demonstrated that by 2050, the proliferation of CAVs could lead to significant reduction in petroleum consumption (90 % reduction) or at the other end of the spectrum lead to large increases in fuel consumption (200 % increase).

Several levers can shift the balance one way or the other. Connectivity and automation can indeed be used for more efficient driving with smarter intersection control and approach and coordinated vehicle-following; this will reduce energy-wasting acceleration/deceleration cycles and will lead to smoother, efficient driving. Platooning reduces aerodynamic drag, which has a significant impact on vehicle energy consumption at highway speeds. Furthermore, there could be opportunities for powertrain design optimization: with more automated driving, acceleration performance will be less necessary, leading to smaller engines, batteries and motors, as well as lighter vehicles. Beyond single vehicle effects, CAVs are likely to change the way the transportation system operates. More intelligent connected and automated driving will reduce traffic congestion, thus allowing for more energy-efficient operations. Automated shuttles could increase the use of public transit, possibly solving last-mile issues, or could become part of the transit system themselves, as shared shuttles. Better knowledge of trip patterns could also alleviate two main barriers for electric vehicle (EV) adoption: short range and long charging times. However, these various levers for increased vehicle and transportation system efficiency are at risk of being negated by a rebound effect: reduced energy savings due to an increase of vehicle-miles traveled (VMT) following mobility improvements, added convenience and potentially lower-cost access due to ride share. If the VMT increase significantly, there will also be a significant increase in energy consumption, despite higher energy efficiency. Increased highway capacity and shorter travel times often lead to more travel—something not uncommon with new roads or highway lane additions. Furthermore, the convenience of fully automated driving may reduce the inconvenience of driving, and may make travelers more accepting of longer travel time and distance. Finally, CAVs may provide additional access to the underserved, which may also produce a substantial increase in total VMT. A holistic research effort is therefore necessary to better quantify the energy impacts of CAVs, to identify the potential for additional energy-saving technologies and eventually to help governments leverage connectivity and automation towards building a more sustainable transportation system.

SMART Mobility research must account for the impact of connected and autonomous vehicle technologies as they become prevalent in the next few decades. The ability to quantify the energy impact of CAVs at the vehicle, corridor, regional and national levels under multiple scenarios (e.g., technology, market penetration, etc.), will be critical in identifying connected vehicle technology that can enhance energy savings and mitigate potential negative impacts. This will allow for informed policy-making and guide further research on connected vehicle technology to maximize sustainability impacts. A key additional outcome will be the identification of CAV-enabled opportunities for greater vehicle electrification, vehicle light-weighting, efficient aerodynamics, powertrain optimization, vehicle utilization, and reduced vehicle miles travelled for further energy use reductions. To this end, foundational work on modeling the energy impact of cooperative adaptive cruise control, field validation of semi-truck platooning, CAV scenario analysis and validation, route based vehicle level control, and urban transportation system models for CAVs has begun and will form the core of the CAV pillar of SMART mobility research.

3 Mobility Decision Science

The potential societal impacts of new transportation technologies on mobility, safety, emissions and energy use are only realized after these technologies are adopted and used in place of conventional technologies. The market adoption and real-world operation of new mobility technologies results from decision-making processes by governments, industry, and ultimately individual consumers. Given the changing travel opportunities offered by new vehicle technologies (electric-drive, connected, automated, etc.) and new mobility-as-a-service options (ride/car sharing), it is imperative to improve the understanding of the decision-making process as it relates to transportation users. These processes, shown in Fig. 2, range in scale from driving decisions (intersection approach, car following, and other driving-style behaviors which greatly influence individual vehicle energy use), to strategic travel decisions (taking a car or bus, leaving earlier to save travel time, where to charge an EV, route choice) [5], all the way to long-term choices (purchase or sale of an electric vehicle, housing location), all of which have significant interdependencies. Connected and automated vehicles have the potential to impact decision making at each of these levels, by improving driving styles, informing and enhancing travel choices, and enabling new vehicle ownership models, for example. Increasingly, the distinction between transportation user and provider is being blurred, with the shift facilitated by transportation network companies like Uber and Lyft. Understanding the behavioral responses of individuals in such situations can lead to better system modeling and design.

One of the largest uncertainties in the development of a future transportation system is how decision-making and behavior will change in reaction to new technologies and system paradigms. As an example, a recent survey has shown wide ranges in willingness to pay for technologies which reduce annual fuel costs by \$600; 10 % of respondents were willing to pay over \$4000, but over 30 % of people were unwilling to pay more than \$500 [6]. Results such as these conflict with what would typically be inferred as rational behavior and underscore the need to understand how to model and incentivize traveler choices at all levels in order to make modern low-carbon and multi-modal transportation systems effective and efficient.

Fig. 2 Research into mobility decision science at multiple decision levels will help to unlock energy reduction opportunities from the deployment of autonomous vehicle technologies

A better comprehension of technology adoption patterns, fundamental human behavior, and decision making processes will lead to a deeper understanding of how technology will impact the transportation system of the future. Likewise, new technology will enable seamless, continuous sensing of transportation related choices, leading to a better understanding of decision-making processes. Understanding the potential range of impacts that advanced vehicle and connectivity technologies will have on the transportation system requires the incorporation of new modeling techniques from cognitive and decision sciences. These disciplines will allow for greater insight into consumer and traveler behavior, through extensive data-gathering, empirical analysis and mobility decision analysis that can be applied towards multi-scale transportation system modeling. Beyond changes in the behavior of individual consumers and travelers, institutional, organizational, and regulatory response to these paradigm shifts will also be of great importance. This is applicable not only to consumers, but also to providers of advanced technology systems that are being incorporated by OEMs into future vehicles. The disruptive potential to a major U.S. industry "eco-system" raises broad concerns since the nation's transportation energy use is largely driven by the platforms offered in the marketplace, and rapidly advancing technology and changing consumer preferences can dramatically alter current market structures.

An opportunity exists to leverage programs, such as DOE's Solar Energy Evolution and Diffusion Studies (SEEDS) initiative and ARPA-E's TRANSNET, as a basis for best practices to expand beyond current research frameworks to develop a better comprehension of technology adoption patterns and fundamental human behavior and decision-making processes. These programs provide a valuable learning opportunity as there is a related focus on understanding the potential adoption (SEEDS) and use of disruptive technology to modifying traveler decision-making (TRANSNET). Mobility decision science research will need to involve the development of a diverse set of analytical, numerical, simulation and conceptual tools and methods for quantitative evaluation; design and implementation of pilot test strategies for modifying current practices; and subsequent assessment of outcomes for impact and scalability. Quantitative evaluation will require multi-scale models—from relatively large national scale to relatively small vehicle subsystem, and everything in-between—of behaviors, decisions, and operations at the vehicle, individual and system levels that can represent the complex interdependencies that exist between each level.

A successful research program in decision science for SMART mobility will result in several key outcomes. Most fundamentally, there should be a greater understanding of individual behavior under future technologies, policies and new transportation system configurations through the application of cognitive behavior science. This will result in enhanced vehicle choice and usage models that will help to accelerate advanced vehicle technology adoption, by enhancing and improving existing models. Beyond vehicle adoption decisions, multi-scale models that can inform more holistic policy decisions, vehicle R&D, and infrastructure investments by iteratively considering individual traveler, corridor-scale, and transportation system impacts will also be developed. The models of individual traveler decisions within the larger transportation system depend on an improved understanding of human behavior and decision-making within the transportation system to enable more efficient travel choices and use patterns. Foundational work in vehicle choice modeling, the development of comprehensive energy/transportation databases, the development of more comprehensive transportation system models, and behavioral analysis of alternative fuel usage is already well underway, and can serve as a starting point for further analysis of connected and automated vehicle impacts.

4 Urban Science

Embedded in any analysis of travelers, vehicles, and transportation and infrastructure should be the study of their interaction with the entirety of the built environment, which we holistically refer to as urban sciences. A focus on the role of urban sciences within the framework of SMART Mobility will (1) provide a platform to integrate traditional transportation research practices with the growing reality that transportation, buildings, grid, water, etc., have cross-cutting impacts and are also becoming more connected and integrated, (2) examine solutions that are greater than the sum of individual parts or systems within a city, and (3) create a mechanism to test, deploy and scale research with local partners. According to the Urban Mobility Report [7], the yearly cost of traveler time and fuel spent in congestion was $121 billion. Today, over 50 % of the global population lives in a city, a figure which is projected to increase to 66 % by 2050, potentially creating even greater pressure on 20th-century transportation systems. Cities emit approximately 70 % of greenhouse gas emissions. On average sprawling cities spend 50 % more on infrastructure per person and households spend $5,000 more on transportation

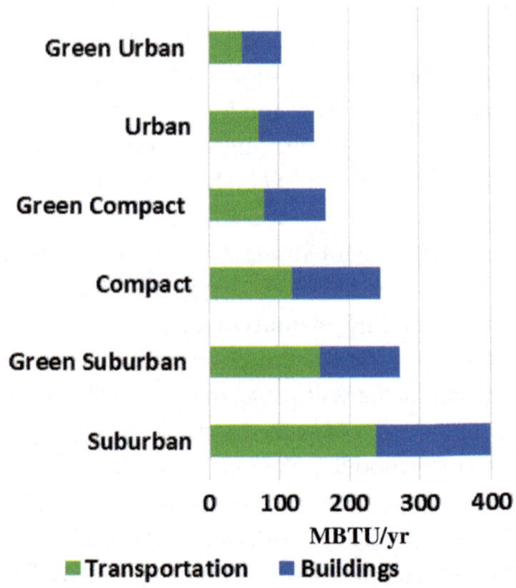

Fig. 3 Transportation energy use (shown in MBTU/yr) depends largely on urban planning and land use

per year. Figure 3 shows how land use and urban planning affect energy use in transportation as well as the opportunities for improving current buildings [8].

Tools are needed that allow us to calculate the energy and resource implication of different urban forms impacted by new transportation technologies. Urban science research will look at integrated, optimized design, planning and operation of large urban areas using computational tools and real-time big data collection. Transportation planning already represents a widespread and robust regional institutional capacity and it will be critical for these entities and the research community to engage. Similar to regional electric grid operators and planners, transportation planners have a significant opportunity to enhance coordination to reduce petroleum consumption and greenhouse gas emissions. The U.S. Department of Transportation's recent report "Beyond Traffic 2045" discusses many crosscutting technological trends that are transforming transportation and can lead to investment and operational decisions that prepare the nation for a future that is energy efficient and livable. Incorporating urban sciences into the SMART Mobility framework will leverage technology evolutions and investments in urban sensing, data availability, and urban infrastructure management systems to better understand and optimize transportation systems.

Increased understanding and sophistication of the urban sciences will enable the development of new city-scale computational models calibrated and validated by large transportation and other city-provided data sets. The frameworks and analytical tools to build and run composite models of urban components related to sustainable transportation (e.g. buildings, vehicle, and utility networks) will allow for more wide-ranging and encompassing analysis of future mobility impacts within the urban system. This is especially significant when considering the interaction between the urban form and new mobility options enabled by CAV technology, driven by changes in user decision-making (e.g., long commutes due to vehicle automation, urban sprawl from increased capacity and travel time). Foundational work in this area has been completed by various local, state, and federal agencies, with a limited focus on energy and environmental implications.

5 Vehicles and Infrastructure

Robust and alternative fueling corridors that both link and function within these urban corridors will also be necessary to meet national energy and environmental objectives. Integrated vehicle-fuel models will explore value propositions (consumer and provider) and opportunities for sustainable transportation deployment to increase petroleum displacement. These opportunities directly tie to emerging vehicle technologies, many of which are now realizing market introduction. These include battery electric vehicles with increasing driving ranges, hydrogen fuel cell vehicles, vehicles capable of operation on advanced biofuel blends and future deployments of wireless electrified vehicle charging (both static and in-motion). These technologies all follow different supply chains and pathways and have implications for infrastructure placement and investment. Coordination analysis and decision-support are needed to help guide the efforts in a way that can maximize alternative fuel vehicles' impact and ease range anxiety.

A series of systematic studies of local and regional alternative transportation corridors can connect existing alternative fueling infrastructure, assist in siting new infrastructure, improve the return on investment of current and future facilities, and potentially enable new business models for transportation as a service. In one study, vehicle characteristics and travel usage patterns helped to prioritize potential EVSE locations from a possible 33,000+ to 281 (as shown in Fig. 4). The economic impacts of this undertaking affect vehicle manufacturers and alternative fuel/EV charging providers and provides benefits over and above local efforts to deploy charging and alternative fueling infrastructure.

A number of states are undertaking their own initiatives to develop a robust alternative fueling and charging network, which presents an opportunity for better coordination. As examples, California has committed $100 million towards new hydrogen fueling infrastructure and the State of Colorado has allocated $30 million for a combination of natural gas, electric, and propane fueling/charging infrastructure.

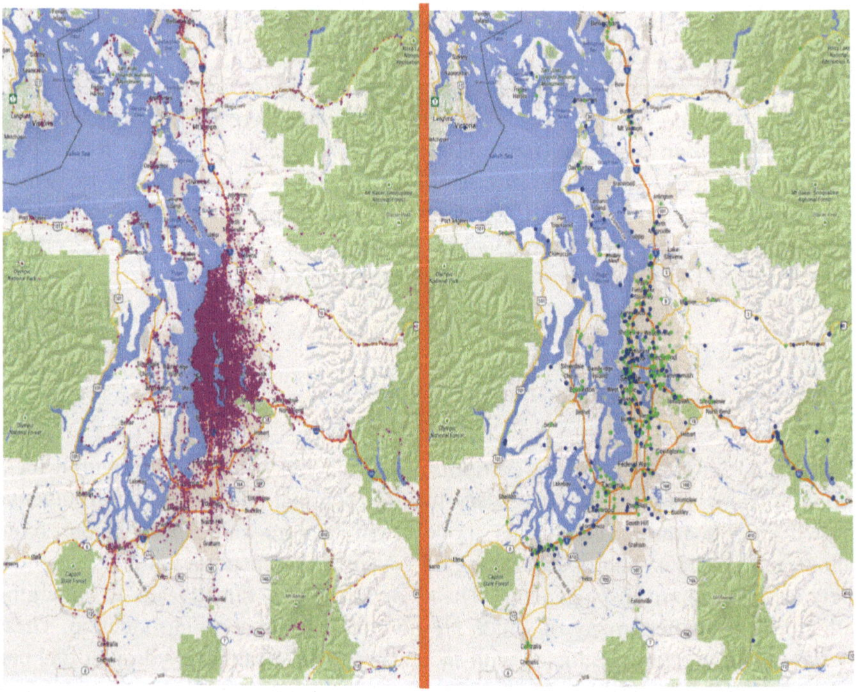

Fig. 4 Combining vehicle characteristics with traveler data helped an effort to prioritize potential EVSE sites in Seattle from 33,000+ locations (*left*) to 281 (*right*) [9]

An integrated model of vehicle-fuel frameworks to explore opportunities for renewables, regional strategies, interactions with the built environment, and the sum total of these effects on consumer choice will be required to fully understand the impact of SMART mobility technologies. This effort will build off of existing foundational work including large-scale data collection effort around EVs and EV charging equipment, the development of demand-based infrastructure siting models, bioenergy feedstock logistics optimization studies, high-octane fuels market research, hydrogen dispenser siting and commercialization, roadway electrification analysis, and vehicle to grid R&D.

6 Multi-modal

Connectivity in automation, both in the vehicle and with the traveler—will enable more accurate, efficient, and timely assignment of modes to match the needs of a specific trip. Creating more efficient multi-modal combinations and connections for moving people and goods from origins-to-destinations, offers an enticing pathway with potential for the U.S. to alter the systemic energy impact landscape—and the issue will become increasingly relevant in out-years. Highly-integrated multi-modal transportation systems enable multiple ways to complete any given trip, whether passenger or freight. Understanding the tradeoffs considered by the private sector when decision-making for various inter-modal trips, and the inter-modal transfer penalties and their tipping points, positions the U.S. to invest in and leverage modal synergies as technologies advance.

Driven primarily by increased demand for goods from economic expansion [10], freight (all modes) currently accounts for about 26 % of petroleum use in

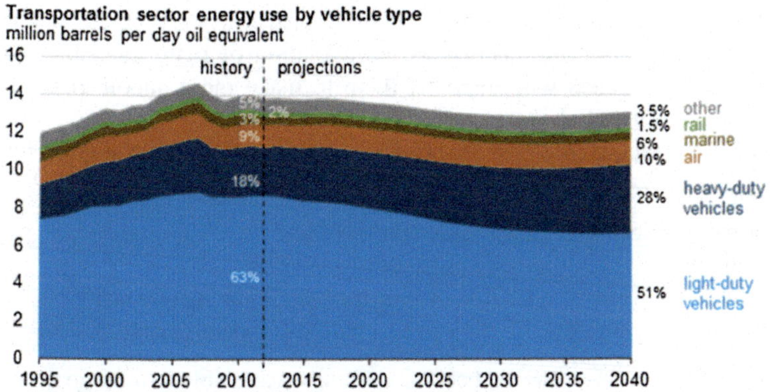

Fig. 5 Energy use share of heavy-duty vehicles is projected to increase 10 % by 2040, while light duty is shown to decrease 12 %. This proposed multi-lab consortium provides an opportunity for the DOE to explore untapped energy-saving opportunities presented by this uptick in freight demand coupled with emerging technologies

transportation with an anticipated increase in demand to over 50 % by 2040—and the share used by heavy duty vehicles is expected to increase from 18 % (2012) to 28 % in 2040 [11] (Fig. 5). Mode-specific energy intensity trends reveal a convergence of passenger mode energy intensities approaching ~2,500–3,500 Btu/passenger-mile, while energy intensities among freight modes vary significantly (ranging up to 21,000 Btu/ton-mile) [12]. The critical need for efficient intermodal freight movement will soon intersect with vehicle automation, as commercial freight-movers are traditional early adopters of technologies (usually fuel-saving) to leverage an economic advantage over competitors. As an illustration, fuel savings from a truck platoon over manually-piloted trucks can range between 2.8 and 9.7 %, which can provide substantial economic benefit for freight companies [4].

Passenger intermodal travel, particularly in urban areas, is equally critical whether by traditional modes of light rail, bus transit and personal vehicles, or by emerging modes such as transportation service companies (e.g. Zipcar, Uber and Lyft), automated transit networks, or eventually automated taxis. As the U.S. experiences greater urbanization, demographic change in the workforce, and new car-for-hire service options, the opportunities for reducing transportation energy use through intermodal-travel will escalate as passengers evolve and adapt to 'mobility as a service' (identified by the Volpe Center in 2014 as a consequence of ever increasing connectivity) [13]. A comprehensive analysis aimed at energy impacts and opportunities in an ever increasing 'mobility on demand' paradigm is thus critical for identifying and implementing measures for long term impact.

Over 86 % of U.S. metropolitan areas offer connected (or nearly-connected) intermodal options—far outdistancing those available in micropolitan and rural environments (small- and medium-sized cities and towns). One collateral benefit to planned expansion of the U.S. telecommunications infrastructure will be a dedicated network for V2X wireless communications and data exchange that offer game-changing improvements to safety and mobility. In parallel, cellular-based data exchange (over ubiquitous smartphones) provides an alternate platform to attain impact in conserving energy, as the evolution of more available real-time information for travelers will empower them to make more unconventional decisions about their mode choice.

Avenues for incorporating more multi-modal efficiencies can take the form of:

- Technology-enabled solutions (e.g. smartphone apps with real-time travel mode information, seamless electronic ticketing, unified multimodal fare payment system, etc.),
- Improved inter-modal connections and transfer infrastructure, and
- Alterations in operational practice (e.g. policies, incentives, etc.).

These pathways are poised to become mutually accessible to a range of travelers and commercial shipper/carriers through new research opportunities that promote higher levels of integration across modes.

An increased understanding of how various modal choices interact in the transport of both goods and people can identify key opportunities for impacting modal energy intensities in the future and increase overall mobility. Research can be particularly impactful in the identification of key combinations for light-duty passenger commuter modes in future urban environments, which can enable greater flexibility and a reduced environmental footprint. Doing so will require the identification and/or development of system models (simulation of urban population centers), optimization methodologies, and emerging technology (both vehicle and infrastructure) that can facilitate and incentivize efficient intermodal choices. Complementary efforts will be needed to quantify the energy impact and potential savings of inter-modal travel as modal transfer barriers diminish from better technological solutions (dynamic information from sensor networks, big data analytics, integrated community design and land use planning, etc.). A natural outcome of this may be enhanced vehicle/transit design to reflect the right-sizing of a trip to more closely match specific passenger needs.

7 Conclusions

Decreasing transportation related GHG emissions will require a significant and multi-pronged effort. This effort is exacerbated by a global demand for transportation energy that is projected to double in the next 30 years [14] as well as the enabling technology for further travel with vehicle automation and connectivity. A vision is needed to reconcile the increasing global demands for transportation energy with a more economically and environmentally sustainable future. A great deal of work is underway to identify more efficient vehicle technologies, and alternative fuels with reduced greenhouse gas footprints, however the underpinning environment in which these technologies exist is at the crux of the proposed concept of SMART Mobility.

This chapter highlights five research "pillars" which must be addressed. Each of these pillars represents either an area in which significant energy reduction could potentially be achieved, or in which greater understanding will allow us to better predict the energy impacts of these new technologies. To date most of the research in these areas has been directed toward improvements in safety and mobility. In order to also capture the potential energy benefit of these technologies research must be directed specifically to understand how these technology work together to provide safety, mobility, and GHG reduction. This will require a coordinated research effort combining the skills and capabilities of government labs, industry, and academia.

This chapter proposes an integrated vision for a sustainable transportation future that focuses on the full transportation system. The approach proposed here builds on the existing opportunity at the nexus of emerging vehicle and fuel technologies and regulatory and competitive pressures. Many important elements of this integrated vision are already incorporated into main research thrusts of key stakeholders' portfolio—from materials that enable light-weight vehicles, to biofuels for low-GHG renewable fuels, and to vehicles operating on renewably generated electricity and hydrogen. By also incorporating SMART Mobility system concepts and approaches into the existing research portfolio transportation energy use and GHG emissions could be significantly reduced.

Acknowledgments This chapter benefits from the expertise and input of the following National Laboratory personnel, many of whom comprise DOE's SMART Mobility Consortium: Kev Adjemian, Idaho National Laboratory; Alexandre Bayen, Lawrence Berkeley National Laboratory; Chris Gearhart, National Energy Renewable Laboratory; Anand Gopal, Lawrence Berkeley National Laboratory; Ron Graves, Oak Ridge National Laboratory; Keith Kahl, Oak Ridge National Laboratory; Eric Rask, Argonne National Laboratory; Aymeric Rousseau, Argonne National Laboratory; Ann Schlenker, Argonne National Laboratory; Alex Schroeder, National Energy Renewable Laboratory; John Smart, Idaho National Laboratory; and Stan Young, National Energy Renewable Laboratory.

References

1. National Laboratory Connected and Automated Vehicle Subgroup (2014) CAVs Energy Impact, presented to the U.S. DOE Transportation Working Group, 17 Dec 2014
2. World Bank (2010) Cities and Climate Change: An Urgent Agenda
3. Intergovernmental Panel on Climate Change (IPCC) (2014) Summary for policymakers. In: Climate change 2014, mitigation of climate change. Contribution of working group iii to the fifth assessment report of the intergovernmental panel on climate change
4. Lammert MP, Duran A, Diez J, Burton K, Nicholson A (2014) Effects of platooning on fuel consumption of class 8 vehicles over a range of speeds, following distances, and mass, SAE 2014-01-2438
5. U.S. Department of Transportation (2014) AERIS Eco-traffic signal timing applications webinar, 29 Jan 2014
6. Singer M. (2015) Consumer views on transportation and advanced vehicle technologies. NREL Technical Report TP-5400-64840
7. Schrank D, Eisele B, Lomax T (2010) TTI's 2012 urban mobility report, Texas A&M Transportation Institute
8. Calthorpe P (2010) Urbanism in the age of climate change
9. Wood E, Burton E, Neubauer J (2015) Measuring the benefits of public chargers and improving infrastructure deployments using advanced simulation tools, National Renewable Energy Laboratory
10. U.S. Energy Information Administration (2014) Annual energy outlook
11. Brown A, Vimmerstedt L (2013) freight transportation demand: energy-efficient scenarios for a low-carbon future. In: Transportation energy futures, national renewable energy laboratory
12. Davis S, Diegel S, Boundy R (2014) Transportation energy data book, 33rd edn. Oak Ridge National Laboratory
13. U.S. Department of Transportation (2014) The smart/connected city and its implications for connected transportation, U.S. DOT, FHWA-JPO-14-148
14. Dulac J (2013) Global transport outlook to 2050, At mobility: technology priorities and strategic urban planning workshop in Espoo, Finland, May 22–23, International Energy Agency

Automated Driving Policy

Bryant Walker Smith

Abstract This chapter summarizes a longer policy paper, *How Governments Can Promote Automated Driving*, which details steps that state and local governments can take now to encourage the development, deployment, and use of automated road vehicles. The chapter has four main parts. *Context* emphasizes the need to think broadly about relevant technologies, impacts, and laws. *Administrative Strategies* identifies steps that governments can take in the course of their ordinary operations. *Legal Strategies* recommends a careful legal audit and provides guidance on the legal changes or clarifications that may flow from such an audit. *Community Strategies* focuses on ways that communities can prepare for and even attract truly driverless systems that are responsive to local needs and opportunities.

Keywords Automated driving · Autonomous driving · Automated vehicles · Autonomous vehicles · Driverless vehicles · Self-driving vehicles · Public policy · Law

1 Introduction

This chapter presents steps that state and local governments can take now to encourage the development, deployment, and use of automated road vehicles. After providing technical and legal context, it briefly describes key administrative, legal, and community strategies. It concludes by urging policymakers to encourage automated driving in part by expecting more from today's drivers and vehicles. A longer policy paper, How Governments Can Promote Automated Driving, explores each of these topics in greater detail and is available at newlypossible.org.

B.W. Smith (✉)
University of South Carolina School of Law, 701 South Main Street, Columbia, SC 29208, USA
e-mail: bryantws@newlypossible.org

G. Meyer and S. Beiker (eds.), *Road Vehicle Automation 3*,
Lecture Notes in Mobility, DOI 10.1007/978-3-319-40503-2_5

2 Context

In examining opportunities for automated vehicles, policymakers should think beyond what a car is and does today. Automation is likely to support and perhaps even require new business cases and use cases. Foremost among these is a shift in transportation from product to service. Instead of driving a car that she owns to work every day, for example, a commuter might instead rely on a private network of vehicles that offer rides on demand. If these rides are shared, such a network may also compete with suburban mass transit systems for customers. At the same time, motor vehicles capable of automated freeway operation over long distances may challenge traditional service operators, such as airlines and hotels.

In broad terms, there are three development pathways toward a fully automated vehicle capable of completely replacing a human driver under all roadway and environmental conditions. Advanced driver assistance systems and automated emergency intervention systems can both be described as "something everywhere" automation that can do only some of the driving—but under many conditions. In contrast, truly driverless systems can be described as "everything somewhere" automation that can do all of the driving—but only under specific conditions.

Existing law may treat each of these technologies and applications differently. Two points are particularly important. First, details matter: New York, for example, uniquely requires a driver to keep at least one hand on the wheel when her vehicle is in motion. But second, broader social context will shape many of those details: Whether a driver acts recklessly by closing her eyes in an automated vehicle may depend on whether the relevant community—including police officers, judges, and juries—deems automation to be "good" or "bad."

New laws may likewise have a range of impacts. Some laws already enacted have prohibited some forms of automated driving, established additional hurdles to testing or deployment, or imposed a superficial structure on existing law that confuses as much as it clarifies. The more meaningful legal review advocated below, however, could identify and then remedy lingering areas of legal uncertainty.

Although this chapter focuses on the public sector, developers of automated systems should contribute to the policy discussion in at least two key ways. First, each developer should detail the specific changes to existing law, if any, that it needs in order to deploy its particular system. Second, each developer should make a "public safety case" that candidly explains how that developer defines reasonable safety, how it satisfies itself that its system is reasonably safe, and how it will continue to do so over the lifetime of that system.

The remainder of this chapter briefly presents potential policy measures in three overlapping categories: administrative, legal, and community.

3 Administrative Strategies

Government agencies and other actors that constitute the bulk of the modern state can encourage automation by preparing themselves, preparing infrastructure, leveraging procurement, and advocating for safety mandates.

Governments should provide their agencies the impetus, authority, and resources to prepare for or even promote automated systems. This includes identifying a single point person for automated driving at each level of government, advancing relevant agency expertise, ensuring that planning processes begin to account for automated driving, and developing break-the-glass plans for responding to early public incidents involving automated systems. These steps will require resources; preparing for automated vehicles involves issues that typically do not confront existing bureaucracies.

Governments should likewise prepare the physical and digital infrastructures that they manage. They should:

1. Prioritize the adequate maintenance of roadways (including pavement conditions and lane markings) to improve the real-life performance of early advanced driver assistance systems;
2. Ensure that policies on the design of transportation infrastructure (including traffic control devices) are clear, consistent across jurisdictions, and actually followed in practice to reduce the frequency with which automated systems must confront unusual roadway conditions;
3. Verify that construction crews and emergency responders follow relevant policies when working on or near active roadways to reduce unanticipated conflicts between automated vehicles and these personnel;
4. Standardize their management of road- and traffic-relevant data to make these data more accessible to digital mapmakers and other potential users;
5. Update existing vehicle registration databases with information about the automation capabilities of every vehicle so that police can readily distinguish between automated and conventional vehicles;
6. Coordinate with national authorities on vehicle-to-vehicle and vehicle-to-infrastructure communications so that this infrastructure is available to those developers that wish to use it;
7. Encourage the deployment of robust wireless communications networks so that developers of automated systems can more reliably share data and updates with these systems after they have been deployed;
8. Make existing congestion management tools (including managed lanes) available for automation-related applications to encourage these applications; and
9. Emphasize neighborhood designs that are consistent with low vehicle speeds to provide roadway environments conducive to early driverless systems.

Governments should also cooperate with each other to increase demand for advanced driver assistance and automated emergency intervention systems by requiring or preferring these systems on vehicles that their agencies, their contractors, and their concessionaires purchase. In addition, state and local governments can push the federal government to move more aggressively in promoting and ultimately requiring more of these safety systems on new vehicles.

4 Legal Strategies

Governments should begin to analyze and as necessary clarify existing law as it applies to automated driving.

A key initial step is to thoroughly audit existing law. This audit should complement the legal analyses that established developers of automated systems should also be expected to conduct. In contrast to the superficial "autonomous driving laws" passed by some states, an audit would attempt to identify every statute and regulation that could pertain to automated driving—including any that might restrict new kinds of vehicles, services, and products. This audit should give particular scrutiny to laws that deviate from actual reasonable practice, consider how enforcement discretion is and should be used to provide more practical flexibility than statutory language might suggest, and evaluate existing legal tools for regulating automated driving.

If the legal audit does identify a need to change or clarify existing law, governments should carefully pursue that change through legislative act, administrative regulation, executive order, legal interpretation, or policy statement. Policymakers should generally seek uniformity, particularly in the underlying legal frameworks that govern vehicles, drivers, driving, insurance, dealerships, and commercial vehicle operations. The use of standardized levels of automation (particularly those developed by SAE International) and the recognition of determinations made by regulators in other states could provide some of that uniformity. Specific changes might include declaring in good faith that automated driving is consistent with relevant conventions on road traffic, exempting the users of automated vehicles from prohibitions on the use of electronic devices, and establishing a clear legal distinction between driver and passenger.

In order to amplify the potential advantages of automated operation, governments should also enforce existing laws related to speeding, texting, driving while intoxicated, wearing a seatbelt, and maintaining a vehicle. Similarly, governments should make vehicle owners and operators bear the true cost of driving by raising fuel taxes, reducing parking subsidies, raising insurance minimums, and allowing or encouraging insurers to implement pay-as-you drive and pay-how-you-drive mechanisms for pricing their consumer products.

More broadly, governments should embrace flexibility by giving agencies the statutory authority to achieve regulatory goals through different means, allowing them to make small-scale exemptions to statutory regimes, and clarifying their enforcement discretion. (Demonstration projects for automated driving within the European Union provide a useful model for these mechanisms.) Many agencies already have considerable authority to encourage or even regulate automated driving, but they need flexibility and resources to appropriately use that authority.

5 Community Strategies

The success of automated driving systems, particularly truly driverless vehicles that are initially restricted geographically, depends in part on how communities react to them. Governments can begin this conversation by thinking locally and preparing publicly.

A community that wants to attract or implement a truly driverless system should demonstrate that it is a strong candidate for such a system by developing a local plan for automated driving. This plan should identify specific needs and opportunities, especially sites such as airports, central business districts, retirement communities, large shopping centers, and areas dependent on last-mile transit routes. Such a plan could inform subsequent proposals to or even stimulate interest from developers of automated systems as well as a variety of state and federal agencies that may have funds available for transportation, community development, energy efficiency, and defense.

Communities should also identify both public and private networks of support for automation. The public network should reach from a state's governor down to local chiefs of police. The private network should involve key interest groups, companies, and even individuals who could advocate for, and possibly collaborate with, developers of driverless systems.

Governments should also begin to understand the broader implications of automation, including but not limited to automated driving. Investing now in structures to manage technology-induced unemployment or underemployment, shifts in land use in cities and within regions, and disruptions in established industries will help the public and private sectors prepare for potentially huge economic and social changes. Although automated vehicles are likely to be only one small part of these changes, these vehicles may also be one of the more prominent symbols of the next technological revolution.

Finally, governments should share the steps they are taking to promote (as well as to anticipate and regulate) automated driving. Knowledgeable points of contact, accurate websites, and ongoing contributions to the broader public discussion will be important in developing sound public policy, attracting initial deployments, building institutional credibility, and appropriately managing public expectations.

6 Conclusion

This chapter has briefly introduced a number of administrative, legal, and community strategies for encouraging automated vehicles. A longer policy paper, *How Governments Can Promote Automated Driving*, further explores these strategies and is available at newlypossible.org.

An important perspective on safety should guide the implementation of these strategies. Governments should appreciate the risks of both automated and conventional motor vehicle travel. Contrary to some assertions, automated vehicles are not yet demonstrably better than human drivers across a full set of driving conditions. Suggesting (and not demonstrating) otherwise risks raising public expectations unrealistically high. At the same time, the considerable dangers of conventional driving are not sufficiently appreciated by the public or addressed by policymakers. In short, the public should be concerned about automated vehicles but terrified about human drivers.

For this reason, governments should expect more from all motor vehicles and their drivers rather than uniquely burdening automated systems. Policymakers concerned about the potential malfunction of automated vehicles should expend at least as much energy on the actual misbehavior of conventional drivers. And policymakers eager to promote automated driving should address subtle subsidies for the ownership and operation of conventional vehicles that could disadvantage new products and services. In other words, governments should encourage automated vehicles by raising the bar for all forms of driving.

How Local Governments Can Plan for Autonomous Vehicles

Lauren Isaac

Abstract Autonomous vehicles have the potential to change all aspects of mobility—from driver safety and insurance liability to car ownership and how Americans commute. It has the potential to disrupt both public and private transportation as we know it. As Google, Uber, the automobile industry, and other organizations continue to make rapid technological advances, it is vital that federal, state, and local governments establish policies, laws and regulations that account for these disruptions. Of utmost importance is finding a balance between guarding public safety while regulating insurance/liability and still encouraging investment in research and development of autonomous vehicles. Most research papers and news reports regarding autonomous vehicles focus on the technological advancement or implications for society (e.g., improved safety and greater mobility for the elderly and disabled people). Many articles cite the importance of government regulation; however, very few provide targeted guidance on how government agencies should respond. The intent of this paper is to outline the role of government in autonomous vehicles and present information local and regional governments need to inform planning and decision-making—both now and in the future.

Keywords Driverless cars · Autonomous · Self-driving · Vehicles · Government · Regulation · Open data · Policy · Privacy · Security · Transportation · Planning

1 Proposed Government Role in Autonomous Vehicles

The following section describes the aspects of autonomous vehicles that will likely require government involvement now or in the future. As shown in the graphic below, the federal government will likely need to update, establish, and enforce policies and regulations around safety, privacy/data sharing, and cyber security, in

L. Isaac (✉)
WSP|Parsons Brinckerhoff, 425 Market Street, Suite 1700
San Francisco, CA 94105, USA
e-mail: isaac@pbworld.com

G. Meyer and S. Beiker (eds.), *Road Vehicle Automation 3*,
Lecture Notes in Mobility, DOI 10.1007/978-3-319-40503-2_6

addition to establishing and enforcing standards. On the other hand, state and local governments will need to update, establish, and enforce policies and plans around mobility, infrastructure, transit, and financials (Fig. 1).

This maintains the government's role of protecting individuals' safety and improving mobility around the country.

1.1 Proposed Federal Government Role in Autonomous Vehicles

It is crucial that the federal government proactively establishes policies and regulations for autonomous vehicles to ensure that passengers and bystanders are safe, but also because it is inefficient, costly, and confusing for all stakeholders to familiarize themselves with multiple, disparate state laws. The following are a range of issues that are most appropriately addressed at the national level:

- *Safety* The federal government should lead the charge in establishing (or updating existing) safety standards for autonomous vehicles, similar to those already in place by the Federal Motor Vehicle Safety Standards and Regulations. Specifically, the federal government could establish standards around manufacturing, vehicle design, infrastructure, and all aspects of data and communications—all with the intent of maintaining safety on our roadways.
- *Privacy/Data Sharing* Because autonomous vehicles will gather a large volume of data to operate most effectively, there are significant concerns about data

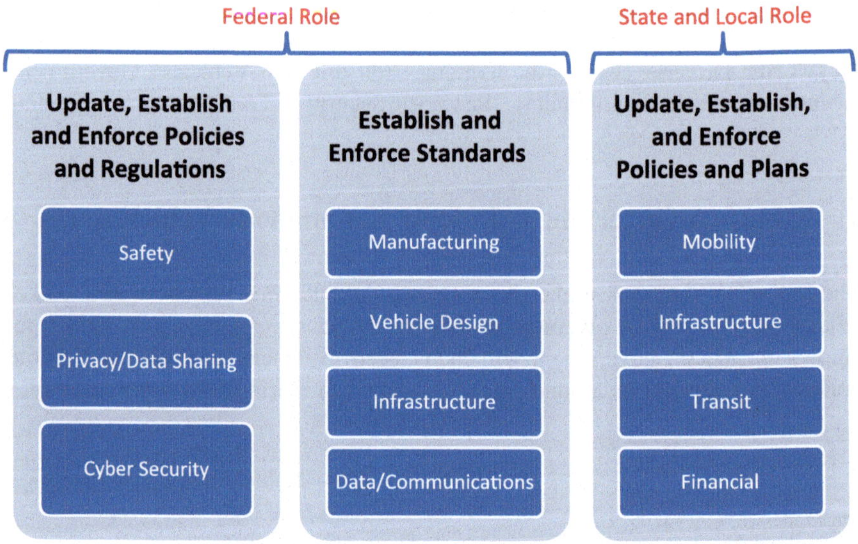

Fig. 1 Proposed Government Roles in Driverless Vehicles

ownership, collection and use. The federal government should, as it has with other data-rich industries, require that the autonomous vehicle industry is secure and transparent with consumers about data ownership, storing, sharing, and security breaches.

- *Cyber Security* Autonomous vehicles could be targets for terrorists, and an attack carries the risk of significant, coordinated traffic disruptions or collisions. The National Institute of Standards and Technology (NIST) is currently developing a framework to improve critical infrastructure cyber security [1] and it is vital that the government ensures that this encompasses the risks associated with autonomous vehicles.

It should be noted that many of the issues described above are relevant for connected vehicles as well.

1.2 Proposed State, Regional, and Local Government Role in Autonomous Vehicles

Autonomous vehicles have the potential to impact states and municipalities in a number of ways: traffic congestion and tax revenues may increase or decrease, current public transit options may need to become more competitive, parking needs may decrease, and roadway infrastructure may need to be adapted (to name a few). Local governments will need to plan for these many changes.

Depending on the governance model utilized in a particular region, different local entities will have jurisdiction over autonomous vehicles. These local, regional, and state government entities may include transit agencies, metropolitan planning organizations, air quality districts, departments of transportation, highway departments, and departments of public works.

The following are a range of issues that are most appropriately addressed at the local level:

- *Mobility* There are many factors that will influence the level of congestion within and around our cities. The level of ride sharing and the competitiveness of public transportation will be significant factors. Additional factors include: increased mobility options for the elderly, disabled, and youth populations, people being willing to live farther from their jobs, and increased road capacity due to shorter headways between vehicles and reduced parking requirements.
- *Infrastructure* Depending on the evolution of autonomous vehicles (and connected vehicle technology as well), local infrastructure will need to keep pace. Specifically, local governments may need to update and reconfigure signage, speed limits, signal timing, roadways and parking spaces.
- *Transit* As autonomous vehicles become more popular, everything from service coverage to vehicle types to labor requirements stands to change. Transit

agencies will need to completely re-think their services, labor needs, and fee structure in order to stay competitive in the new transportation environment.
- *Financials* The widespread use of autonomous vehicles will have potentially significant financial consequences for local governments. Taxes, parking fees, speeding tickets, parking real estate, and incident management costs are just a few of the government revenues and costs likely to be impacted. Moreover, the local government may need to identify new sources of revenue to pay for infrastructure (similar to today). Local governments should understand the impact of autonomous vehicles ahead of time and prepare accordingly.

The following section describes specific actions the state, regional, and local governments can take both now and in the future to proactively address the issues likely to arise from the proliferation of autonomous vehicles in our society.

2 Proposed State, Regional, and Local Government Role in Autonomous Vehicles

Autonomous vehicles will have significant impacts on many aspects of society, and, as such, local, regional, and state governments need to start planning for these now. In fact, local governments need to consider the following planning and policy actions now and in the next decade, especially since the AV roll-out is well within transportation planning time horizons:

- Near-Term Planning Activities
- Medium- to Long-Term Activities, including planning, infrastructure modifications
- Policy Activities

While short-term planning activities need not be time or labor-intensive, the policy activities and medium- to long-term recommendations will likely require significantly more resources and political will. This is described in more detail in this section.

2.1 Near-Term Planning Activities

The next few years will likely be a time for continued development and testing of the autonomous technology. The recommendations outlined for the near-term are focused on supporting the advancement of the technology and positioning the government to successfully plan for the future autonomous vehicle society.

2.1.1 Support Testing Activities

States will likely continue to be responsible for autonomous vehicles' licensing and testing requirements in order to ensure public safety. This includes establishing the standard for who can "drive" (or be responsible for) an autonomous vehicle, and how and where it must be tested. This paper does not have any specific recommendations for these requirements; however, consistency between states and collaboration with the technology developers is highly encouraged.

Additionally, government entities may consider offering government-owned closed-campus land parcels as locations for technology developers' testing. Examples include: college campuses (e.g., Santa Clara University), islands, and former military bases (e.g., GoMentum). This can support the advancement of the technology while also keeping government informed and connected to its private industry partners.

2.1.2 Stay Educated on Autonomous Vehicle Progress

It is vital that local, regional and state governments become educated on the state of this constantly-evolving industry. Government representatives should follow autonomous vehicle developments—both in technology advancement and national policy development, in the United States and internationally. These are a few approaches to monitoring these developments:

- Subscribe to ITS America Smartbrief and attend local and national ITS conferences
- Subscribe to www.Autonomoustransportation.com listserv
- Subscribe to the Association for Unmanned Vehicle System's International (AUVSI's) listserv and attend their conferences
- Set up a weekly news alert for "Autonomous vehicles" and "driverless cars"
- Follow blogs for emerging developments in AV, such as Driving towards Driverless [2] and Driverless Car Market Watch [3].

2.1.3 Establish Communications and/or Coalition with Autonomous Vehicle Stakeholders

Ideally, government entities will become partners with this ever-growing community. Stakeholders to consider can include local representatives from automakers, technology developers, insurance brokers, privacy advocates, attorneys, and even the general public. Participation in a coalition can be a way to garner or provide valuable input as policies and plans are developed in future years.

2.1.4 Establish Policies and Plans with Consideration for the Future

In addition to increasing awareness around autonomous vehicle technology, local, regional and state governments should develop current plans and policies with an eye to the future and a focus on safety. The ability to create plans and policies that are flexible and easily updated will be vital as technologies evolve and society's needs change. For example, a transit agency's fleet management plan may specify vehicle needs for the next 30 years; however, it should also acknowledge the potential for a significant change in vehicle technologies and vehicle sizes. This acknowledgement will be important to factor into the agency's planning functions and vehicle procurement strategy.

2.1.5 Encourage Open Data Sharing

As more information becomes available, the government is in a position to encourage the open sharing of data. While it is important to preserve people's privacy, open, anonymized data can improve government decision-making and help the government to develop more informed policies and plans. Information on what open data is and why it's important can be found here [4].

Data philanthropy, or the willingness of private companies to share their data, may not happen. In order to motivate these private companies, government may consider giving companies strong tax incentives for sharing data for public good. Additionally, public and private companies will need to work together to identify models for data sharing in ways that respect personal privacy and security and enable companies to do well by doing good.

2.2 *Medium to Long-Term Recommendations*

The following recommendations present activities that will likely need to happen irrespective of which AV scenario occurs. These include planning activities, infrastructure modifications, and a few additional miscellaneous activities.

2.2.1 Planning Activities

The following activities represent short and long-range planning considerations that will continue to evolve over the next few decades. It will be important to re-evaluate all of these activities as autonomous vehicles become more pervasive in society.

1. *Update travel demand model* As more information around autonomous vehicles and their usage becomes available, travel demand models will need to be updated. The travel demand models should ideally reflect updated information

regarding where people are living and working, how many trips they are taking, what level of shared rides are occurring, and the vehicle ownership model. It should also capture any changes associated with freight delivery. All of these factors are likely to impact travel behavior. Modeling these impacts will likely be refined as the technology is developed further; however, this research paper describes one approach to updating an existing activity-based travel model [5].

2. *Evaluate Road Capacity Needs* Based on findings from the travel demand model, update long-range plans to reflect new road capacity needs. Road capacity requirements may increase or decrease depending on whether or not VMT increases or decreases and whether or not vehicle throughput is increased.
3. *Assess transit service delivery plans and fleet requirements* Local government will need to determine how they can most appropriately work with private mobility providers to ensure equitable, fairly-priced mobility options for everyone. This will require the local transit agency to coordinate with all mobility stakeholders (e.g., travel demand modelers, private mobility companies, taxi providers, paratransit providers, etc.) to determine the appropriate level and location of transit service.

   The transit agency will also need to re-evaluate its fleet management plan in order to incorporate autonomous vehicles in its fleet. This will have significant implications for labor requirements (and Union agreements), maintenance facilities, maintenance workers, safety and security of passengers, etc.

   Note: This applies to public school bus systems as well.
4. *Forecast financial implications* Utilizing a cross-functional group of stakeholders, government officials should examine every line item of the budget to evaluate the potential financial implications of autonomous vehicles. Examples of line items to consider are:

   - Parking revenues
   - Speed ticket violation fees
   - Tax revenues related to vehicle purchases, registration fees, and VMT
   - Transit agency costs and revenues
   - Incident management costs
   - Insurance costs
   - Government fleet transition to autonomous vehicles
   - New enforcement activities
   - Unemployment insurance

2.2.2 Infrastructure Modifications

The following activities highlight the many infrastructure changes likely necessary over the next couple of decades. Many of these details will not be determined until the autonomous vehicle technology is more fully developed and, in some cases, until these vehicles are prevalent in society.

1. *Designate/certify roads for AV and/or manual operation* Designating or certifying roads implies that the roadway owner (a government entity) would provide some framework for evaluating and confirming a roadway is acceptable for a specified usage. It is debatable whether or not this will be needed. Some see this as inevitable since roadway owners will assume some level of responsibility for the "driving public." On the other hand, others think certifying roadways (whether for autonomous or manual cars) would be too resource-intensive and would also imply a higher level of liability than is appropriate. That being said, examples of certification requirements may include clear markings for the vehicles to read, markings, or signage to designate the road as certified, appropriate signage regarding speed limits and construction zones, and communications to other road users about the vehicle types allowed on that roadway.
2. *Update traffic signs and markings* Local governments should monitor and provide feedback on any updates to the Manual on Uniform Traffic Control Devices (MUTCD) [6] and local versions of the manual. This will provide information regarding changes to standards regarding how traffic signs, road surface markings, and signals are designed, installed, and used.
3. *Reduce lane width* Assuming lanes are marked appropriately, autonomous vehicles will not require the traditional 10–12 foot lane widths on local roads or highways. This may not be a necessity, but the reduction could increase the capacity of roadways, provide added space for bike lanes, and/or improve walk-ability. Note: this could have positive impacts for cities undertaking Vision Zero initiatives.
4. *Alter speed limits* Autonomous vehicles will travel at or below the speed limit specified on roads. In a fully autonomous vehicle society, it may be just as safe to have higher speed limits, especially on highways. It is likely that the methods for setting speed limits will change over time. Federal guidance regarding speed management safety can be found here [7].
5. *Adjust traffic signal locations and timing* A fully autonomous vehicle society will likely introduce entirely new travel patterns. As a result, the local government may need to alter the traffic signal locations and timing. In the future, traffic signals may be unnecessary or they may be highly adaptive to current traffic flow (Note: This introduces new concerns regarding pedestrian and biker safety). Additionally, local governments may consider prioritizing transit and shared occupancy vehicles at intersections.
6. *Eliminate/Reduce parking and add more drop-off/pick-up locations* Many parking spots (both on and off-street) may be unnecessary (or re-located) due to lower private vehicle ownership and the potential for autonomous vehicles to park themselves in remote locations. For these reasons, more infrastructure for passenger pick-up and drop-off locations may be required. Additionally, parking space sizes may be reduced.

7. *Develop new predictive models for pavement maintenance* The timing for pavement maintenance may be quite different from today's requirements. Level of roadway usage may significantly increase or decrease (depending on whether or not VMT increases or decreases) and autonomous vehicles may be lighter and operate in a way that has less impact on the roadway.

2.2.3 Miscellaneous

The following activities do not fit into any of the previous categories. They include updating the local government's enforcement and incident management functions.

1. *Update enforcement function within government* Existing enforcement activities, including the pursuit of speed limit evaders and drunk drivers, will likely be completely unnecessary in an autonomous vehicle society. In fact, the enforcement requirements will likely shift to include the following types of activities:
    - Ensuring cars are certified and meet government standards
    - Certifying roadways and ensuring autonomous vehicles are driving safely on these roadways
    - Ensuring safety and payment of transit passengers
    - Ensuring any managed roadways (Express Lanes, HOT lanes, etc.) are collecting appropriate revenues
    - Ensuring data is shared appropriately
2. *Update incident management function within government* Currently, the government spends a significant amount on responding to car accidents. While some level of car accidents will be inevitable, the number of incidents is forecasted to decrease substantially. As such, the local government can reduce that function and potentially reduce costs.
3. *Update government workforce to match needs* It will be necessary for governments at all levels to ensure that its workforce has the skill sets needed to fill different positions and functions that emerge as people change their travel behaviors and patterns. While this is hard to predict at this point, it will be important to follow autonomous technology developments around the world and invest in training throughout this time of significant change. In the short-term, governments should consider hiring people who can actively support the many changes likely to occur in the government during this time.

2.2.4 Policy Activities

Autonomous vehicles have the potential to greatly increase or decrease VMT depending on the level of ride sharing and vehicle sharing that occurs. The following activities represent policies that the government can put in place to influence how autonomous vehicles can affect VMT, urban sprawl, and/or parking requirements. These policies can be put in place at any time; however, the sooner they are in place, the sooner any potential negative impacts of autonomous vehicles can be mitigated.

1. *Update roadway policies and infrastructure* to manage the VMT impact Autonomous vehicles present the risk of people increasing the amount that they travel significantly. With the propagation of electric vehicles, the cost of fuel likely decreasing, and the ability to be productive while in their car, people may travel more often and for longer distances. For that reason, the following changes to the roadways may help to dis-incentivize this potential increase in VMT:

   - Adopt or increase roadway tolls in general and/or specifically for single occupancy vehicles
   - Add or designate more high occupancy vehicle ("HOV"), high occupancy toll ("HOT"), and express lanes
   - Add congestion pricing in and around urban areas or downtown cores/central business districts

2. *Adjust land use polices to reduce urban sprawl* Autonomous vehicles present the risk of people being willing to live much farther away from their workplaces since their commute times may be newly productive. For that reason, it will be important for local governments to establish policies that encourage high density, walkable communities in order to minimize urban sprawl. Examples include:

   - Add more transit-oriented development and transit service, in general
   - Maximize areas within walking distance of each other by minimizing space allocated to parking
   - Create policies and processes that encourage developers to build walkable communities
   - Create and enforce urban growth boundaries
   - Support infill development
   - Develop policies that make greenfield development very expensive and onerous (which currently describes trying to build infill development)
   - Improve the quality of the schools in the urban core
   - Add "green-belt" type development buffers to constrain sprawl
   - Limit septic-based development (AB)

3. *Adjust the tax/fee structure to dis-incentivize car ownership and/or parking* Autonomous vehicles may result in an increase in single occupancy vehicle trips (and, ultimately, increased VMT). Taxes and fees can be increased or decreased to incentivize sharing rides and dis-incentivizing private car ownership and single-occupancy vehicle rides. Examples of taxes that can do this include [8]:

- Sales tax on private vehicle purchases
- Tax on vehicles miles travelled
- High fees for public parking and high taxes for private parking
- Reduced (or even subsidized) costs and parking fees for shared ride services
- Reduced (or even subsidized) costs for bike share, shuttles, and other first/last mile solutions
- Increase local "guaranteed ride home" programs so commuters feel confident they can get home irrespective of how they got to work

4. *Alter parking policies to reduce the need for private parking* If consumers own their own autonomous vehicles and rarely share them, the parking needs will probably remain similar to today. On the other hand, increased ride and vehicle sharing could significantly reduce the parking requirements. Parking policies can be established to minimize the land dedicated to parking and manage where parking spaces are located. Examples include:

   - Eliminate minimum parking requirements in zoning laws
   - Restrict or limit the number of parking spaces allowed in residential developments (and reduced even further if along transit corridor) (and encourage/require those spots to be dedicated to car sharing providers)
   - Design parking facilities in such a way that they can be re-purposed at a later time
   - Require developers to develop parking management plans that outline how parking requirements can be minimized
   - Developers must pay for the right to develop parking spaces and the government can use that funding to pay for parking in designated (and possibly remote) locations
   - Establish a city-wide parking space cap
   - Dedicate parking space to car sharing companies or shared vehicles. Additionally, car sharing companies may have exemptions for parking time limits or unlimited access to street parking
   - Institute variable priced parking to proactively manage how parking spaces are used.

5. *Change transit pricing* As described earlier in this section, transit agencies will need to determine the appropriate level and location of transit service. For example, smaller vehicles with dynamic service may make public transit competitive with private vehicles and other mobility providers. In conjunction with this planning effort, it will be important to evaluate transit's "competition" and ensure the transit service is competitively priced. Transit has always been subsidized and this will likely need to continue—especially if it is intended to support the low income population and potentially low ridership routes.

3 Conclusion

Autonomous vehicles are coming, with or without government involvement, and there is no question that they will have a significant impact on society. People may choose to live farther from their jobs, roads may become even more congested, and public transportation may no longer exist. On the other hand, road safety may be drastically improved and current non-drivers may have new mobility options. The autonomous vehicle could, literally, change how we approach all aspects of life.

Moreover, the lines between public and private responsibilities with regards to mobility are blurring. Automakers are becoming service providers, the taxi business model is becoming obsolete, and ride-hailing companies are providing new mobility options. This has created new "competition" for public transit and, as such, is requiring the government to re-think its role in mobility.

With the coming of autonomous vehicles, the United States government (at all levels) has the opportunity to proactively establish regulations, policies, and plans that can continue to support the autonomous vehicle revolution while keeping the traveling public safe and providing a positive example for governments around the world.

References

1. Enotrans.org. http://www.enotrans.org/wp-content/uploads/wpsc/downloadables/AV-paper.pdf. Accessed 4 Aug 2014
2. Driving Towards Driverless Cars. http://drivingtowardsdriverless.com/. Accessed 3 March 2016
3. Driverless car market watch. http://www.driverless-future.com/. Accessed 3 March 2016
4. GovTech.com. http://www.govtech.com/data/Got-Data-Make-it-Open-Data-with-These-Tips.html?page=2. Accessed 2 Feb 2016
5. Childress S (2014) Using an active-based model to explore possible impacts of automated driving. Submitted for presentation at the Transportation Research Board 2015 Annual Meeting, Washington, D.C. http://psrc.github.io/attachments/2014/TRB-2015-Automated-Vehicles-Rev2.pdf
6. FHWA.com. http://mutcd.fhwa.dot.gov/. Accessed 4 April 2015
7. FHWA.com. http://safety.fhwa.dot.gov/speedmgt/. Accessed 4 April 2015
8. Shareable.net. http://www.shareable.net/blog/policies-for-shareable-cities-transportation. Accessed 3 Aug 2015

Part II
Human Factors and Challenges

Shifting Paradigms and Conceptual Frameworks for Automated Driving

Patrice Reilhac, Nick Millett and Katharina Hottelart

Abstract This article takes seriously how discourses of automated driving shape the world we are designing and the public's perception of it. The metaphors which organize our thought and scaffold our conceptual frameworks betray our point of view: legacy, engineering-based or user-centered. Valeo's Intuitive Driving strategy centers on the user experience and therefore on the evolving relationship between user and technology. Relationships are built on trust. Automated Driving involves an evolution of agency in a high stakes context with new implications for trust generation. It also entails a shift which has a huge impact both on industry and user: from horsepower to data-power. This shift fundamentally alters the nature of the relationship between human and vehicle. In the design of this relationship anthropomorphism is a central issue. Building a trusting human-machine relationship in automated driving inevitably means dealing with social robotics and affective computing where anthropomorphism in technology has been explored for many years. But the specificity of the automatic driving moment must be attended to: this is the only robot with an interior private mobile space. This new being will need a specific behavior designed for it, and already, a new discourse to speak of it.

Keywords Intuitive · Intuitive driving · Driving automation · Human robot relation-ship

---

P. Reilhac (✉) · K. Hottelart (✉)
Valeo Schalter und Sensoren GmbH, Laiernstr. 12, 74321 Bietigheim-Bissingen
Germany
e-mail: patrice.reilhac@valeo.com

K. Hottelart
e-mail: katharina.hottelart@valeo.com

N. Millett
NMA Branding and Management Consultancy, 212 Boulevard de Charonne,
75020 Paris, France
e-mail: nick@nma.eu

G. Meyer and S. Beiker (eds.), *Road Vehicle Automation 3*,
Lecture Notes in Mobility, DOI 10.1007/978-3-319-40503-2_7

1 Etymologies of Driving Automation

It is no exaggeration to describe the rise of automated driving as a paradigm change. With every new paradigm our whole conceptual framework needs to be reorganized. Initially the industry looks from an engineering point of view at what is technically possible and then tries to materially realize that possibility. The car is focused on and then the driver is added in. In fact, more than anything, "end-user" could be said to describe this addition of the driver as an afterthought at the "end". One automated component is quickly added to another, with both eyes fixed on the growing agency of the machine, and an odd glance over to the human in the equation.

Placing the user centre-stage is at the heart of Valeo's *Intuitive Driving* strategy. This is why we pay particular attention to how we speak about what doesn't exist yet. The world of automotive automation is taking shape for the public through the language the industry uses. As yet it is more a semantic world than a concrete existent world. This world *itself* will have agency and will hinder or facilitate the adoption of automated driving technologies. This is why we would like to pause here, take a step back, and explore what is at stake in how we frame the discourse of automated driving.

The obvious place to start is at the top: "automated driving". We choose to use this phrase against the prevalent industry term of "autonomous driving". There is a world of difference between "automation" and "autonomy": the first says "self-acting" and the second says "self-governing". We should not take such a distinction lightly. At a time when we should be easing a new technology into the public's consciousness with care, a time when highly respected public figures and opinion leaders gift the media the opportunity for scaremongering about "rogue AI" [1], it is not exactly reassuring that the industry itself promotes the image of cars as autonomous: i.e. which govern themselves, are free, independent and self-ruling. On the one hand we all refer to a steady, sound, incremental increase in automation, and on the other we name this *process* by its end state, and implicitly, a higher level of being: one that industry leaders have no hesitation in invoking triumphantly as "Superhuman" [2]. Moreover, as we shall see, the autonomous vehicle which it gives its name to the process will more than likely never exist.

This future fantasy of total self-sufficiency is repeated in the public's preferred terms for automated driving [3]: "self-driving" and "driverless". Once again, the human is already out of the loop, and the car is able to act *totally* alone at all times. In these three versions of autonomy, the car-as-thing replaces the human and in so doing gains what we might call a *strong agency* which encourages an anthropomorphism of *self* and *will*, with its concomitant dangers. It is no exaggeration to call this dominant imagery "Googlespeak", since it is they who are driving the conversation of autonomous driving in the English-speaking world (Figs. 1 and 2).

"Automated driving" leads us to a very different place and set of imagery. In fact, if you take "driving" in its original sense of "impulsion", then you find that the notion of "self-driving" is purely tautologous to that of "automobile". Originally "automobile" was a French adjective, as in "véhicule automobile", describing a

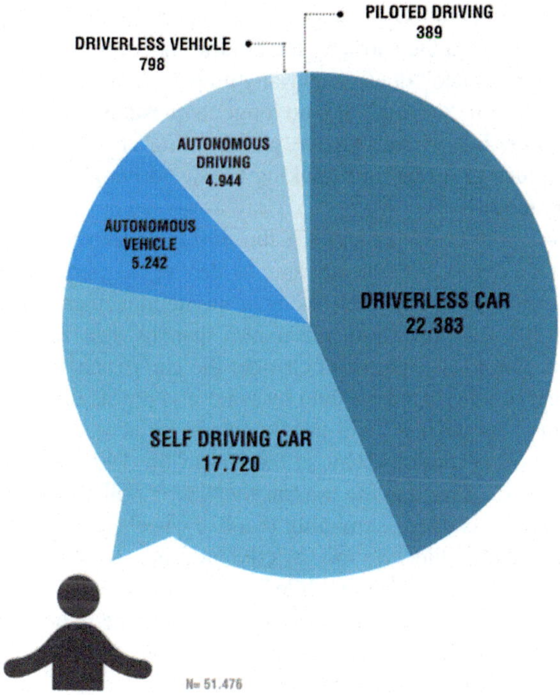

Fig. 1 How do users refer to automated driving? [3]

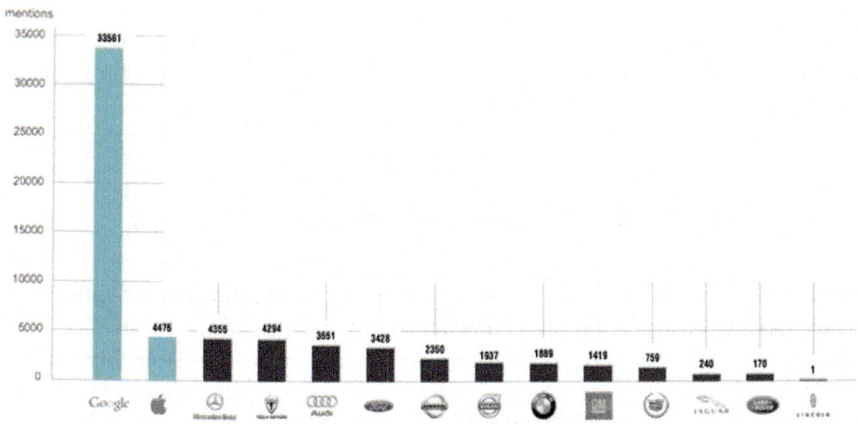

Fig. 2 What are the major brands people talk about in the context of autonomous driving? [3]

carriage which has internalized its horses from a horse-drawn carriage where what you “drive” is external to the carriage: the horses (its “engine”). So semantically at least, “self-driving vehicle” doesn’t add anything to “automobile vehicle”, unless you update the sense of “driving”. It’s no longer a question of the vehicle having its own motive power, but of that motive power being able to activate itself. This etymology continues to format our thinking today: “driving” is still primarily seen as a powertrain activity.

What is obscured in these semantics is the shift from *horsepower* to *data-power*. There is a concomitant shift in emphasis to driving as *perceptual* activity. The automated car’s potential for smooth and appropriate movement will depend less on its engine capacity and powertrain mechanics than its data acquisition and management powers. This is a significant change: the car’s capacity for action will be determined more by *what it knows* than by what it is physically capable of. And as we shall see, this “knowledge” will not be owned by it.

No doubt one of the reasons why, at the end of the 19th century, “automobile” won out against its rival for naming the car, the “locomobile”, was this specification of difference from the carriage. Anything that has wheels, and so can change place, can in fact be a “locomobile”. So too, strictly speaking, anything with a motor and wheels could be called an “automobile” today. Perhaps it is time for a new semantics and with it a new taxonomy of machines and a new ontology of the automotive sector. Perhaps it makes sense to talk of an evolution from *automobile* to *automatomobile* to *autonomobile.*

However, we would want to continue problematizing the notion of autonomy and so would tentatively suggest that if you add the primacy of *data-power* to the *automatomobile* what you get next might be a *robomobile*. As we move out of the Age of the Automobile and into the Age of the Robots this type of suggestion is more than idle play. It places us in the contexts of users’ evolving relationship to technology and of *data-power*; together they may just totally transform the automotive industry within the next decade.

2 Growing Trust: The Levels of Driving Automation

As we have seen, the logic for preferring the description “automated driving” to the other variants is that it doesn’t immediately confront us with the finality of an ultimate stage. Furthermore users are *familiar* with automation as something in their world which has different degrees and contexts. If we care about shaping the perception of the future of driving, and if we care about fostering user adoption of new technology, then it is important to carefully craft the conceptual framework of this world we are creating.

The industry has a more or less standard conceptualization of the evolution of driving automation. The well-known table from Stanford summarizes the dominant description of levels and includes the equivalents from the German BASt and the American NHTSA [4, 5].

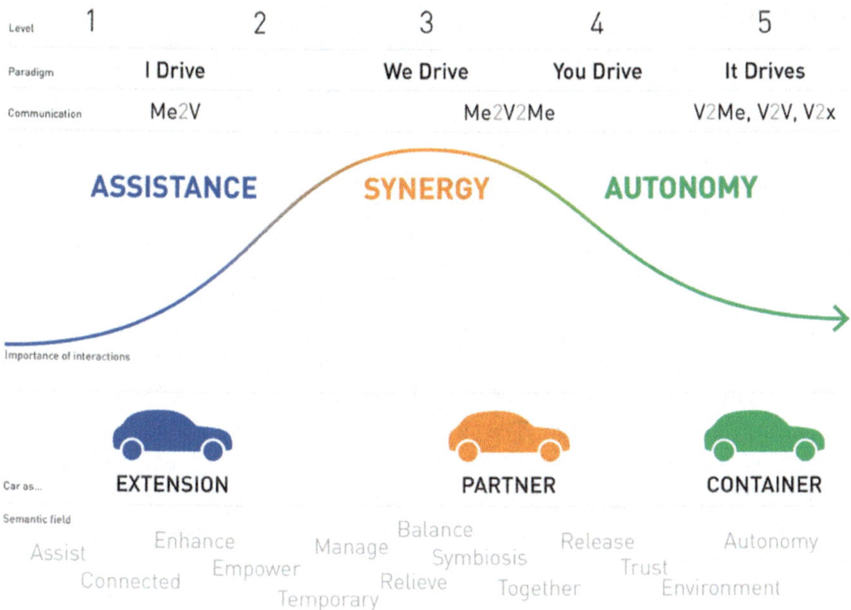

Fig. 3 Automation level table

We use a simplified version of this table to try to conceptualize the evolution of automated driving in terms of users' experience (Fig. 3).

We identify three *changes in nature* in user perception occurring in the five *degrees* of the SAE scale. The first of these paradigm shifts is between "I Drive" and "WeDrive" [6].

A new paradigm forces a new language. In 2011 Valeo's approach to driving assistance evolved from *Smart Driving* to *Intuitive Driving*. With Advanced Driving Assistance Systems the emphasis shifted from an object-oriented engineering process to a relationship-oriented experience design. For users systems' sensory and relational capacities are more significant than their cleverness. They perceive certain features of ADAS as intuitive and automation as the natural evolution of these features. At the level of simple assistance (Level 1), IDrive expresses the fluidity of human-machine interaction involved in moments of delegation or information provision.

At the more advanced levels of partial automation (Level 2) we already begin to sense the emergence of the next paradigm of WeDrive. The categories of the old paradigm start to become strained until at Level 3 (conditional automation) the discrete categories of driver and passenger break down: the Driver can also be a Passenger—a Driver-Passenger, or what we call a *Drivenger*. With the new paradigm a whole world of new needs and possibilities emerges. *Time* is introduced into the relationship between driver and car as they exchange periods of driving. The Valeo cockpit concept *Valeo Mobius*® is designed for this new world where the act

Fig. 4 Möbius strip

of driving is shared between driver and car, and where the driver gains the new power of being a passenger too.

Valeo Mobius® figures the primary feature of this driving world: An extended HMI which enables a fluid, continuous exchange of agency between human and vehicle, and the safe integration of users' digital experience to facilitate the hand-over of control [7]. The hyphen in "human-machine interaction" is expressed in the topological figure of the Möbius strip which subverts our usual Euclidean way of representing space and relation, and which helps us start to adapt to the demands of the new paradigm. The strip appears to have two sides but really only has one: what is discrete is actually continuous. The only way of discovering this is by taking the time to go over the whole strip. Where one becomes two and two becomes one again is imperceptible, and this describes the intuitiveness of the interaction (Fig. 4).

The regulative ideal of WeDrive is a relationship of continuous intuitive communication between *drivenger* and car. Combined agencies and multimodal realities will increasingly structure the reality of our digital moment; from the perspective of the old paradigm the new situations may well appear as paradoxes. This is part of the power of the Möbius strip figure: it expresses the limits of the Euclidean paradigm. The intuitiveness sought in WeDrive is probably the closest driving will come to the classic metaphor of the horse and rider [8]. However, because this metaphor elides the fact of the rider's necessary skill-set, perhaps the most useful analogy to express the 2-in-1 synergistic *experience* of WeDrive as a stage of Intuitive Driving is that of theatre training's Mirror Exercise where two participants face each other and follow each other's movements until neither they nor observers know who is leading any more: two individuals in one symbiotic movement.

Like the Möbius strip, this analogy places both terms on the same level such that they become reversible, unlike in the horse metaphor. The *relationship* which forms at the centre of the user experience is one of reciprocity. This symmetry is important to move us away from the unilateral instrumentality which is encoded in the notion of "Human Machine Interface". We replace this notion with that of a "Human Robot Relationship" (HRR) in order to meet the demands of the new paradigm and advance towards the following one: introducing dual agency, time and reciprocity into the concept. This new concept reminds us that we are not designing *things* but *relational user experiences* [9] involving trust.

As soon as we start evoking reciprocity and symmetry in the context of trust we are faced with the question of anthropomorphism. If we follow our analogy with the theatre game, then the mirroring of the smartphone in the HMI is only the beginning. *Mirroring* becomes a key way in which the "machine" interrelates with the *drivenger* and behaves *as if* it understands the human. We move from the HMI question of how the human interacts with the machine to that of how the machine interacts with the human. The next step is to design these interactions so that they form a relationship. The emphasis is no longer placed on "intuition" as "ease of use" but as "ease of relating": the HRR is ideally about human and machine intuitively *understanding* each other. Before technological "smartness" reduced "intelligence" to the purely cognitive, "intelligence" in its human connotation actually might have included the *non-cognitive* [10] (we should mention in passing that "non-cognitive" is one of the synonyms of "intuitive"). At the level of WeDrive, a *partnership* needs to be formed out of the building blocks of relational understanding: *intuition, intention* and *expression.*

The leaders in the industry who set the agenda for the development of the future of driving try to understand this new relationship of partnership via metaphor. The favorite at the moment appears to be that of the car as animal. No longer the horse, however. Now it is the domestic animal, and in particular the paradigm of the pet: the dog. In 2008 Carlos Ghosn, CEO of Nissan, was at pains to stress that the dashboard robot in the Pivo concept car was not a toy but a way to help driver and vehicle "bond" as "they would with a pet" [11]. While today Daimler takes the metaphor even further: "Our mission: to breed better 'dogs'" [12].

The dog or: "man's best friend". It's a highly emotional metaphor, if not sentimental: the automated car as pet animal. And it's being used at a time when machines are *literally* replacing pets in the form of "companion robots". We can glimpse an evolving fuzzy syllogism: automated cars are like pets, pets are replaced by robots, therefore automated cars are robots. If this is the case and if the HRR is as central as we claim it to be, then is the automated car less a *robomobile* as we suggested earlier and more of a *carbot*? In other words, as *data-power* takes over from *horsepower*, and as the HRR takes over from the HMI, should an automotive design philosophy centered on the user experience not increasingly see itself more as a subset of *social robotics* and *affective computing* than of automotive engineering?

This question is driven by several issues:

1. The "self" or subjectivity of the driving machine as it is being constructed by industry discourse and by Google, as we discussed in our opening section on etymology [13].
2. The rise of data-power enabling new players expert in user experience design to enter the market: like Apple and Google.
3. Social robotics having the most expertise in relationship-building between human and machine. And much of the experience in this field has been developed by working with *vulnerable* users: children, people with disabilities and the elderly. In other words, publics where the stakes are high and gaining trust is a key issue. Arguably, Google is doing social robotics as much as it is doing automotive engineering.
4. The consumer robot market is growing seven times faster than the manufacturing robot market [14]. By the time fully automated driving arrives, robots will be part of users' lives.

For very different reasons, two of the most promising target groups for automated driving, millennial and the silver market, are also two of the hardest to market automated driving to *and* key targets for social robotics. The evolutions we are highlighting suggest that the question of trust in automotive will be mediated by users' new interaction with technology via the robotic ecosystem. Discovering where in the levels of automation changes of nature occur for users and therefore where trust is a determining factor is crucial. The industry needs to take advantage of the developments of social robotics and affective computing, to identify *where these developments are user led,* to examine *how* trust is generated and what relevance this might have for driving user experience. Finally, it will be necessary to explore how *cultural difference* determines attitudes to automation. We already know that the speed of adoption of automated driving will be much faster in countries which have favorable regulatory systems and well-developed robotic cultures.

3 Growing Trust: Anthropomorphism

Social robotics' pets, or "companion robots", are mostly humanoid robots like Aldebaran Robotics' Pepper, Blue Frog Robotics' Buddy and Dr. Cynthia Breazeal's Jibo. This last announces itself, like a pet, as "one of the family". The degree of human likeness decreases from Pepper to Jibo. In the former likeness is predominantly via physical appearance and in the latter via physical behaviour (humanlike head movements). Social robotics works with *anthropomorphism* in two ways:

1. In full understanding of the so-called *Eliza Effect* [15]: the human tendency to read human qualities into technological artefacts.
2. Designing features that trigger anthropomorphizing and elicit emotion. *Faciality*-giving some degree of face to robots—is central to their approach.

As we mentioned above, anthropomorphizing metaphors spread fast in automated driving as soon as we start replacing "machine" by "robot" since for many a robot just *is* a humanoid machine but especially since *agency* itself is seen to be a primarily human property. What interests us here is whether the anthropomorphic is just a bridging metaphor to an unknown future how far anthropomorphizing is likely to develop in automated driving as opposed to social robotics, and what type of anthropomorphism should be sought, if at all other ways of growing user-machine trust in the automotive sphere.

New technologies take getting used to. This is why the incremental advance to autonomous driving needs managing carefully. The choice of metaphors in such a process is important. An example can be found in users' perception of "driverless trains". In a user study [22], 93 and 72 % of respondents thought that a "fake" driver room should be present on a driverless train. They needed to behave *as if* the train was being driven, regardless of whether it was or not; *as if* a human was still in control. In the automotive case, no doubt the "as if human" design adds a layer of

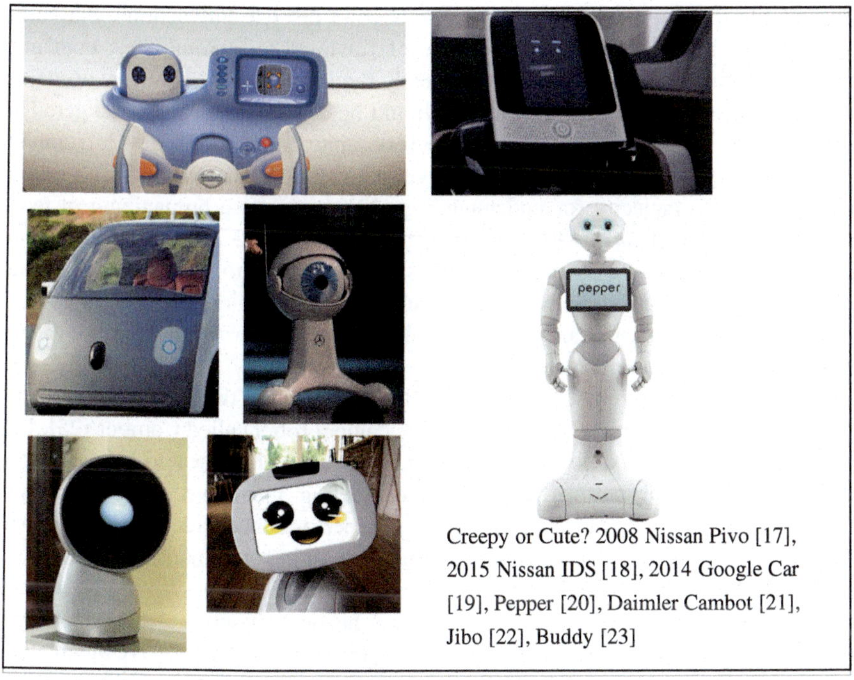

Fig. 5 Various robots: 2008 Nissan Pivo [16], 2015 Nissan IDS [16], 2014 Google Car [17], Pepper [18], Daimler Cambot [19], Jibo [20], Buddy [21]

reassurance: the alien machine can be seen to be *well-intentioned* and *harmless*. This is very similar to social robotics' use of human likeness and what we call "the cuteness factor": robots made to look like *little* animals or humans seem more friendly, docile and "user friendly". In Nissan's 2015 IDS concept the "face" of the 2008 Pivo concept car's dashboard robot has been sublimated into the automated mode's oversized smartphone style console with two blinking eyes on its screen (Interestingly, this console appears to have no function whatever other than anthropomorphism and to communicate its "trusted partner" status.) (Fig. 5).

Of course *faciality* has always been a design issue for cars. Inferring personality traits from car "faces" is another example of humans' anthropomorphizing tendencies. Google unashamedly went for the toy face approach in their 20014 self-driving car design. And as the 2015 Nissan IDS concept shows, we are likely to enter into a new phase of anthropomorphic design as automated cars gain new powers of *expression* to communicate with other cars and people outside. But this is also where we start to leave social robotics behind, or at least where we have to develop our own *automotive* social robotics or *social* automated driving: the car is *two-faced* and entertains two HRRs, inner and outer. Our primary interest in this paper is in the inner HRR which distinguishes the automotive from social robotics *stricto sensu*.

In the WeDrive paradigm of Intuitive Driving the car gains an agency which needs to dovetail with the *drivenger's* agency. The fluidity of this dovetailing requires the development of a relationship beyond discrete interactions. Playfully, we can talk about "automating" as a type of "mating" for automobiles: both intimate coupling and befriending. Concretely and minimally, the car needs to understand when the driver is, wants or should be driver and when passenger. The *faciality* that the social and automotive robot interfaces above develop are an attempt to physically produce a humanlike subjectivity that will make us immediately infer this capacity for understanding [23]. But does a humanizing of technology mean that it has to literally and physically become *humanlike* [24]? The Eliza Effect suggests that as the car demonstrates its ability to "read" situations and our intentions, we will quickly attribute human characteristics to it anyway.

The "cuteness factor" can prime and enhance users' perception, bringing "enjoyment" and "love" into the relationship with technology as Daimler CEO Dr. Dieter Zetsche exposed at CES in 2015 when introducing the F 015 autonomous driving concept by conversing with a toylike camera robot called Cambot [25]. But in the automotive context physical human likeness takes a back seat to *behavior*. It is the car's behavioral performance of understanding the *drivenger's* intentions, expressing its own and translating them into safe performance which will drive trust. The Valeo Mobius® HMI concept already builds trust in performance by smoothing the driving hand-over process and providing an interface for non-driving activity: it understands that when driving the *drivenger* needs to continue passenger-style activities, and when a passenger the *drivenger* needs to be ready to drive again. This is an evidential form of trust-building through positive experience of performance.

"Cuteness" on the other hand responds to a combination of *dispositional* trust (culture, age, gender, personal traits) and *initial, learned* trust (pre-existing knowledge) in the form of an a priori judgement. As we try to work out the psychodynamics of WeDrive, the Valeo Mobius® HMI concentrates on *situational* trust (setting, difficulty, task, risk), and *dynamic, learned* trust (system performance, reliability, validity, errors) [26]. Dispositional and initial trust will impact the design of the HRR and in particular its range of adaptability to different *drivengers*. As it adapts apparently more and more intuitively, the car will be attributed a human-like being without the need for *physical* likeness. The success of this attribution and its integration into a seamless relationship will depend on a combination of behavioural and experience design. The requirements of this design and the way trust is constructed will evolve further as WeDrive mutates into YouDrive.

4 From WeDrive to YouDrive: Adaptive and Distributed HMI

At the high automation of Level 4 YouDrive introduces another new dimension of time into the driving relationship. The *drivenger* now has swathes of time where no supervision or monitoring of the driving activity is necessary. During these periods distraction caused by NDR activities no longer needs to be minimized [27] and so they will become more ample, as will the possibility for the *drivenger* to participate more in the social space of the vehicle. This new temporal condition creates a new relationship between *drivenger* and car which requires a different kind of bond to the playing partnership of WeDrive. Now the relative autonomy of the two parties means they are attached above all by trust alone: the *drivenger* needs to be able to look away, to close their eyes even: to delegate totally. This is a considerable intensification of the HRR where a functioning synergy needs to evolve into complicity. The content of the trust relationship is now less the reinforcing evidentialism of physical, operative interactions. It is a more fluid, intangible construct closer to "a meeting of minds".

Concretely the evolution at YouDrive involves the following:

- The *drivenger* needs to spend time apart from the driving function of the car: the interior hardware need no longer be for driving. At least temporarily there need no longer be a cockpit. The HMI should adapt itself to the activity in process.
- Since the *drivenger's* attention may be far from the driving console and, in any case, since the console now may have been transformed for another activity, the modality for recalling the driver to take over the driving function needs rethinking. The HMI now needs to be distributed throughout the interior of the cabin. Whether auditory, haptic or visual will depend on what the *drivenger* is doing at the time: the interior of the car (and not just the console) needs to "sense" and "know" in detail what the *drivenger* is doing.
- The interior architecture is released from a rigid configuration centered on the driver and driving position. The interior space and its hardware becomes labile, metamorphosing between automated and manual modes.

User tests carried out with the Valeo Mobius® HMI module have shown that already at WeDrive, there is a need for a more ambient, pre-symbolic medium of feedback expression. At Levels 4–5, we imagine that this medium will be even more continuous, ambient and angled towards thin, peripheral awareness. In short: the interior cabin as a whole will need to gain its own sensorium and its means of expression will need to be distributed throughout too. The intuition of Intuitive Driving now becomes more of a perceived capacity of the car and the intuitive relationship less centered on a functional, physical task. The intuitiveness of WeDrive and YouDrive is now being constructed via the car's capacity to understand the *drivenger's* intentions and to express its own. In Intuitive Driving there is no time for thinking: the instantaneous reactivity of the car is less a demonstration of intelligence than of *aliveness* (Fig. 6).

Agency is inflected in a new way at YouDrive: the car is trusted with vehicle operation for long periods and in diverse situations, it is trusted with the metacognitive capacity of knowing when it cannot deal with a situation and of communicating this in time to the *drivenger*, wherever they might be looking. The *Monitoring* which helped develop the HRR at WeDrive takes on a new power here. As is often the case a technical industry term gets so worn during concept development that it becomes but a contentless placeholder when it finally gets to meet the

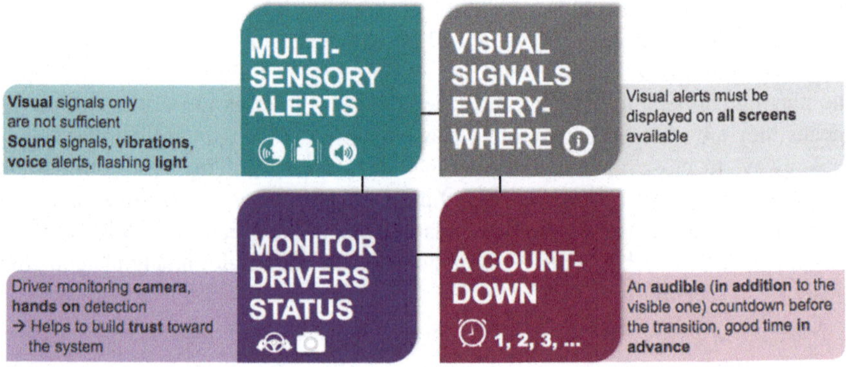

Fig. 6 Key elements for a smooth transition from automated to manual driving (Valeo)

user. So "monitoring" masks the far-ranging implications for the user experience represented by the becoming-sensitive of the car's interior which occurs in advanced visual and physiological monitoring. To make automated driving feasible it has been necessary to concentrate on the construction of a sensorium for external perception. Now a sensorium needs to be constructed for internal perception too: this allows it to relate to all the occupants in its space, to physically reconfigure itself and to respond to the occupants in a multitude of ways. The "face" of the HRR will be everywhere and nowhere, but if we could see it we will *feel* that its expression would be one of *caring* and *concern*.

This vehicle with its double sensorium is an extraordinarily different being to what today we call a "car". But it is also very different to any of the robots that exist. It is not only that finally the space of the car comes into its own. It is that it becomes a *living* space in both senses of the term: a space for living and not just being transported in, and a space that seems to be alive. But also a space with which we are permanently in *physical contact*. There is a parallel here with the home automation company Nest's desire to move away from the notion of the Smart Home to that of the Conscious Home [28]. Smartness belongs to *tools;* intuition and consciousness, if only metaphorically for the moment, belong to *relationships*. We maintain that the illusion of "life" is more important than the fact of smartness insofar as intelligence is conceived of as traditional computational intelligence.

This leads us to underline two aspects of the automotive HRR which need to be factored into the evolution of users' perception:

1. The automotive robot is not just a machine endowed with AI and movement, it is also a space in which we dwell. This can make it simultaneously more threatening (we are not just faced with a potential rogue element, we are inside it!) and more welcoming.
2. The car will be capable of machine learning. Functioning by inductive inference rather than by deductive reasoning it will seem to have humanlike mental capacities. It is possible that humans will have some sort of pedagogic role. In

any case, the HRR will be predicated on a process which will already be encouraging complicity and empathy. *Mental* anthropomorphism will be an inevitable component of the user experience.

We want to emphasize the singularity of this new assemblage in part to disrupt any metaphors that might initially help us bridge the gap to the future only then to hinder our progress. In YouDrive, trust will be generated not only by the car's capacities for driving but also by its capacities for relating to the occupants. Emotional and social intelligence, simulated or not, conveyed through intention and expression, will be central. Logically the car's status as *container of space* and locus of non-driving becomes more important than its status as transporter. But experientially this "container" is much more than that: It is a *living responsive environment.*

5 Living Responsive Environments and the You of "YouDrive"

When the *drivenger* can get in the car, turn to "it" and say, as if to their partner, "You drive!", and then turn over and close their eyes, we will have reached Level 4 of the SAE automation scale: high automation. Our hypothesis is that at this moment the HRR will have been developed to the point that users really will interact with their automotive environments *as if* they were people. This does not mean that the car will now have the "autonomy" of a person nor that the human will be out of the loop, and it certainly does not mean that the vehicle will legal and ethical status of "person".

There is a scale of automation whose highest tenth level is described in this way: "The computer decides everything, acts autonomously, ignoring the human" [29]. This could be the credo of some luddite dystopian rogue automobile nightmare! Indifference to the human is certainly encoded in the current hegemony of "autonomous driving", and the efforts to which Google has to go to soften their car's image is a symptom of this. We can certainly do all that we can to ensure that the highest level of automotive automation will be user-centered. In fact, we believe that the shift from *horsepower* to *data-power* and the release of both the driver and interior space from functionality makes the car more than ever *our* habitat.

Technology is in the process of animating our inanimate habitats. Objects of all types gain sensoriums, communication organs and means of expression, without changing form or function. They maintain relationships with each other and with humans. In this context the fully automated car, sensing, communicating and expressing itself, internally and externally, will retain its singularity in the robotic ecosystem even only by the fact of its being a *mobile space*. It is too early to speak with the barest of certainty about the full automation of Level 5. It cannot be known *how* and *when* or *if* the robotic in the automotive sphere will cohere in one direction: as robot smart-*phone*, or embedded in environments, or embodied in humanoid form, or embedded in humans. But since our bias goes towards ubiquitous, invisible and embedded, we can propose the following hypotheses:

1. The increasing rarity of driving by humans will increase the aesthetic character of the driven space. “Aesthetic” in all senses of the term: a place for expression and recreation; a place for sensory experience and social sharing of that experience; a place which senses. Continuous adaptation and materials innovation for expression will increase the organicism of this increasingly alive responsive environment.
2. In the YouDrive paradigm hardware reconfiguration will be developed as a function of driving mode. In the new paradigm of Level 5 full automation the plasticity of the driven space and its architecture will be increased, probably along two vectors:
   - Shared Driving: purpose-built designs. Sharable or rentable fleets of different types of driven spaces: work-oriented, play-oriented, sleep-oriented etc. The *drivenger* becomes pure passenger.
   - Owned Driving: reconfigurable, modular interiors. The steering wheel may remain as an option. But the *drivenger* will be an endangered species; a sort of retro-futurist mobile homer perhaps.
3. Level 5 will, at an advanced stage, see the car able to drive without any human occupant inside. But it will still only be autonomous *from* human *drivers*. It will not be autonomous *in itself*. Initially the human will probably exist as a teleoperator safety net, much like an aeronautic control tower. The relationship will be “cold” because the vehicle will be almost totally networked, almost traveling on digital tracks. This will likely be a new paradigm which we call “ItDrives” where the vehicle is totally networked and the “it” refers less to the vehicle itself than to the overall network. Notions of trust, freedom, performance and the HRR will need reconceptualizing in this paradigm. Data will totally replace the powertrain as the core of driving; transport space will be completely striated digitally and vehicles will circulate on this digital network. As a result it is more than possible that the “being” or “self” of the car, like the *drivenger* before it, will take on new properties and powers. Perhaps it will become a *purely* social robot.

In terms of the act of driving the best topological figure for advanced YouDrive and early ItDrives is probably the borromean knot which expresses the interdependence of human teleoperator, vehicle and network. But at an advanced level of ItDrives even the human teleoperator will disappear and the network will, in effect, do the driving. We would gradually return to the dual agency of the Möbius Strip, only with the *drivenger* replaced by the Network, and then to the unitary agency of ItDrives, just like IDrive, but with Network *ego* replacing human *ego*. At this time Humans will probably only ever “drive” in VR simulations, and it will be all the more pleasurable and “real” than any driving as we understand it today could be in the context of a future where the Network holds agency and the car itself is actually *less* autonomous than ever. In a sense absolute “autonomous driving” (from the human) is where the *car* has the *least* autonomy (Fig. 7).

It is hard to imagine what the anthropomorphic status of ItDrives will be. With the new perspective of this endpoint of Automated Driving, we can however be sure

Fig. 7 Borromean knot

that by the time even WeDrive happens there will be no need for *visual* anthropomorphic features on cars. The cuteness factor may be useful for prepping the public's imagination now, but the ubiquity of robots will mean that trust will emerge less superficially in the automotive sphere. The bridging metaphor of humanness will only end up hindering development. Better to already start trying to grasp, express and develop the ontological specificity of the automated car, the singular experiences, modes of existence and design spaces of possibility it gives birth to.

To do this we need to *already* be exploring how to construct the "You" of YouDrive in terms of the HRR. There are many ways of producing this subjectivity other than *faciality*. As we have seen, the Eliza Effect shows how *expressive behaviour*, even of the most disembodied type, is enough for us to infer analogous humanlike being in machines [15]. Once the car's HRR software gets its inductive inferential machine up and running and engages with its human occupants, complicity will form fast, as we reciprocally engage with its perceived mental capacities: insofar as those capacities demonstrate *social* and *emotional* intelligence. A name and a voice will no doubt help, but further physical anthropomorphism will generate little more trust [30]. The automotive "You" can be constructed through its understanding of *intentions* and its *particular* mode of expression which takes into account its physical singularity (it is a living responsive environment for humans) to invent a robotic assemblage which does more than just seek to replicate humanness *since we know that in order to relate humans will anyway inject as much humanness as they can into artefacts* [9].

6 Conclusion

In conclusion, we can point to how further user research is necessary to be able to construct a more detailed automation scale that is meaningful from users' point of view. Where are the paradigm shifts for them? Are these emotional? Or psychological? What new properties and capacities emerge at each level? Do they care about the logical and engineering logical changes in nature? What consequences do the mismatches have for designing future automotive experience? And *how* confident can we be that what the user thinks or imagines *now* will be pertinent once the shift happens? The research will need to predict how their perception will evolve as their relationships with technology evolve. More than ever the automotive is part of a wider *data-driven* world which shapes user expectations.

On the cusps of paradigm shifts research needs to be innovative and to build conceptual frameworks through interaction with other disciplines. In particular, interfacing with the Arts, Humanities and Social Sciences is required not only because of the complex ethics that machine agency introduces into the industry, but also to develop a new ontology of automated agents which can inform theories of the new assemblages of users and automated automobile entities. As we drop our mechanistic habits of the HMI days and develop HRRs, it will become clearer that we will need not just user interface strategies but clearly defined philosophical positions [9]. Out of this research and this thinking a discourse can be consciously built with which the industry can itself drive the public conversation on automated driving and already start conceiving it *with* and *in* users' imaginations. This discourse will move beyond the rhetoric of "superhumanness" and "autonomous driving' to construct a new grammar and syntax expressive of the relational specificity of automated driving. It will imagine the automated car less as an accumulation of ADAS features and more holistically as a world of HRRs, alive *enough* to be much more than the sum of its parts.

References

1. Hawking S (2015) Elon musk and various other signatories' public manifesto from January 2015, https://en.wikipedia.org/wiki/Open_Letter_on_Artificial_Intelligence
2. Dr. Dieter Zetsche's keynote at CES in 2015, Daimler CEO
3. Hyve Science Labs: See Autonomous Driving—The User Perspective by, TUHH and INSIUS (October 2015), the authors research user language by online data mining. http://www.hyvescience.net/wp-content/uploads/sites/4/2015/10/autonomous-driving-report.pdf
4. Shladover SE, Lappin J, Denaro RP, Smit BW (2014) Introduction: the transport research board's 2013 workshop on road vehicle automation. In: Meyer G, Beiker S (eds) Road Vehicle Automation. Springer
5. Stanford.edu http://cyberlaw.stanford.edu/blog/2013/12/sae-levels-driving-automation. Accessed 14 Oct 2015
6. Reilhac P, Millett N, Hottelart K (2015) Thinking intuitive driving automation. In: Meyer G, Beiker S (eds) Road Vehicle Automation 2. Springer
7. Diederichs F, Bischoff S, Widlroither H, Reilhac P, Hottelart K, Kaiser F (2015) Smartphone integration & level 3 car automation—new cockpit concept & evaluation in simulator
8. Flemisch FO, Adams CA, Conway SR, Goodrich KH, Palmer MT, Schutte PC (2003) The h-metaphor as a guideline for vehicle automation and interaction, NASA/TM-2003-212672
9. Reilhac P, From "human factors" to "User Experience" for level 3 car automation, the 2-in-1 Mobius concept, ERTRAC Annual Conference, Brussels. http://www.ertrac.org/uploads/documents_publications/2015%20Conference%20presentations/Valeo.pdf. Accessed 05 Oct 2015
10. Turkle S (2011) Alone Together. Basic Books, New York
11. Kaufmann A (2007) Nissan's creepy dashboard robot could make production. http://www.motorauthority.com/news/1025865_nissans-creepy-dashboard-robot-could-make-production

12. Herrtwich RG (2014) The promises and pitfalls of vehicle automation, daimler, automated vehicles symposium presentation, San Francisco 2014
13. Barr A, Ramsey M (2015) Google tries to make its cars drive more like humans. http://www.wsj.com/articles/google-tries-to-make-its-cars-drive-more-like-humans-1443463523
14. Hicks J (2015) Age of companion robots: jibo, pepper and now buddy. http://www.forbes.com/sites/jenniferhicks/2015/04/25/age-of-companion-robots-jibo-pepper-and-now-buddy/
15. Hofstadter D (1995) Preface 4: the ineradicable eliza effect and its dangers in fluid concepts and creative analogies: computer models of the fundamental mechanisms of thought. Basic Books, New York
16. 2008 Nissan Pivo: Image © Nissan
17. 2014 Google Car: Image © Google
18. Pepper. http://www.theverge.com/2014/6/5/5781628/softbank-announces-pepper-robot
19. Daimler Cambot. http://mercedesblog.com/cambot-real-star-las-vegas/
20. Jibo: https://www.youtube.com/watch?v=3N1Q8oFpX1Y
21. Buddy. https://futurscience.wordpress.com/2016/02/23/serie-sur-les-robots-francais-buddy-le-premier-robot-familial-francais-est-ne/
22. Fraszczyk A, Brown P, Duan S (2015) Public perception of driverless trains. Urban Rail Transit 1(2): 78–86
23. Deleuze G, Guattari F (1988) A thousand plateaus, trans. In: Massumi B, Athlone 1988. London
24. Reilhac P, Millett N, Hottelart K (2015) Thinking intuitive driving automation. In: Meyer G, Beiker S (eds) Road Vehicle Automation 2, Springer
25. Luchian E (2015) Cambot, The Real Star in Las Vegas, January 2015. http://mercedesblog.com/cambot-real-star-las-vegas/
26. Hoff KA, Bashir M (2015) Trust in automation: integrating empirical evidence on factors that influence trust. Hum Factors 57(3):407–434
27. Diederichs F, Bischoff S, Widlroither H, Reilhac P, Hottelart K, Moizard J (2015) New HMI concept for an intuitive automated driving experience and enhanced transitions. In: 7th conference on automotive user interfaces and interactive vehicular interactions 2015 workshop proceedings, 2015
28. Constine J, Nest's CEO calls its tech the "conscious home" because "it's not smarter than you". http://techcrunch.com/2014/11/19/conscious-home/. Accessed Nov 2014
29. Wickens CD, Hollands JG, Banbury S, Raja Parasuraman R (2013) Engineering psychology & human performance 382
30. Waytz A, Heafner J, Epley N (2014) The mind in the machine: anthropomorphism increases trust in an autonomous vehicle. J Exp Soc Psychol 52:113–117

Truck Automation: Testing and Trusting the Virtual Driver

Steven Underwood, Daniel Bartz, Alex Kade and Mark Crawford

Abstract This chapter addresses the testing and evaluation of the virtual truck driver. While the primary focus of the discussion is on verification and validation in model-based systems engineering it also touches upon testing for certification, establishing regulations, public investment, and research and development. A reference architecture for automated driving coordinates designs at the vehicle and system levels for increased interoperability among components and improved efficiency. A model-based systems engineering approach exploits automated vehicle systems domain models as a primary means of information exchange to help manage the complexity and provide analytical support for efficient architecting, design, verification, and validation. These models support the testing and evaluation process for functional safety design and certification. Finally, demonstration pilots, operational testing, and natural use testing, combined with system design artifacts, are critical to public and regulatory acceptance of the virtual driver. Although safety must be assured, the primary challenge is how to make such assurances without relying on a human driver and vouching for the virtual driver under all allowable driving situations and conditions. This chapter provides some ideas on how all of this might come together and help bring fully automated vehicles to the market.

S. Underwood (✉)
Connected Vehicle Proving Center (CVPC), University of Michigan-Dearborn, 4901 Evergreen Road, Dearborn, MI 48128-1491, USA
e-mail: underw@umich.edu

D. Bartz
SAE Reference Architecture and Interfaces (RAI) Task Force, San Francisco, CA, USA
e-mail: danielbartz@gmail.com; danielbartz@autonomation.net

A. Kade
Ground Vehicle Robotics, US Army TARDEC, 6501 East 11 Mile Rd., AMSRD-TAR-R/264, Bldg.200C Rm. 1130C, Warren, MI 48397-5000, USA
e-mail: alex.kade.civ@mail.mil

M. Crawford
Research and Advanced Engineering, Ford Motor Company, Building 2, 20000 Rotunda Drive, Dearborn, MI 48124, USA
e-mail: markcraw@umich.edu

G. Meyer and S. Beiker (eds.), *Road Vehicle Automation 3*,
Lecture Notes in Mobility, DOI 10.1007/978-3-319-40503-2_8

Keywords Automotive · Trucking · Trucks · Fleets · Testing · Evaluation · Automation · Architecture · Verification · Validation · SysML · Safety · Army · Driving system · Simulation · Systems engineering · Functional requirements · Model-based · Reference architecture · Interfaces · Certification · Standards · Vehicle · Pilots · Operational testing

1 Introduction

Perhaps the easiest way to envision the automated truck driving system is to imagine what it takes to be an excellent truck driver. Among other things the excellent driver should be in good health and be perceptive, responsive and adept at maneuvering the truck and trailer in its immediate environment and in accordance with the rules of the road. Their senses are attuned to the state of the truck. The excellent driver knows safe stopping and following distances, and turning radii with and without the trailer under different conditions and loads. They keep the vehicle in good working order, up to code and to the extent possible, out of harm's way. This driver is also skilled at avoiding or steering-clear of road hazards including pedestrians, objects and vehicle events that could potentially cause crashes. And when things go wrong the driver is able to bring the vehicle to safely to a stop.

While this discussion about the excellent truck driver is not exhaustive it provides some insight into what is expected of the automated driving system and how it should behave in order to keep the passengers and cargo safe. It is much more than a collision warning or crash avoidance system; it drives the vehicle within selected bounds from beginning to end. For those in transportation safety it might suggest adding a new column for the virtual driver and a new row for near crash behaviors to the Haddon Matrix addressing pre through post-crash dynamics [1].

However, while Haddon focuses on the crash event, automated driving systems will address all aspects of human driving as well as new capabilities like short gap platooning and multi-vehicle coordination. The virtual driver has the potential to increase safety directly through more attentive perception and responsive handling of events, and indirectly by navigating at safe speeds, distances, and gaps that human drivers often neglect. The virtual truck driver may be able to follow a lead vehicle with a short time gap with greater safely and reliability than a human driver. These safe and extended driver behaviors are system level targets for testing, approval, and in many cases certification for automated trucking. The goal is to perform these behaviors more comfortably and safely time-after-time over years and across miles with fewer errors and with better performance and reliability than even excellent human drivers.

Testing and evaluation will help ensure a proper design the excellent virtual driver and build public trust as well as legislative and regulatory backing. System testing will also be necessary to assure consumer acceptance and responsible manufacturing and maintenance of self-driving trucks. Over the long term testing will help to improve design, grow trust, and future certification of automated trucks.

2 Autonomous Mobility Appliqué System

United States Army Tank Automotive Research, Development and Engineering Center (TARDEC) is taking an incremental approach to automation by integrating mature sensors and control systems into U.S. Army and Marine Corps tactical vehicles to assist drivers and enable future autonomous operation. This project and related architecture and test activities will be used to illustrate some of the test and evaluation themes addressed in the chapter.

The Autonomous Mobility Appliqué System (AMAS) will reduce the dangers of driving in combat by providing soldiers with Active Safety and Warning systems while at the same time providing as platform for incremental adoption of automated vehicle systems. The first phases of AMAS deployment will take the base platform and add semi-autonomous convoy control, relieving the soldier of the driving task, reducing fatigue, eliminating rear end collisions. Additionally these technologies have been proven to enhance operator situational awareness, to enable a more effective response to hostile situations. Future increments of AMAS will provide autonomously navigation through urban and rural environments and negotiation of pedestrians, intersections, and oncoming traffic. The TARDEC plans follow the following sequence of increasing functionality: (1) driver assist and active safety, (2) leader/follower behavior, and (3) full automation.

The full AMAS system consists of an integrated By-Wire/Active Safety Kit and an incremental Autonomy Kit. AMAS uses a combination of automotive and specialized sensing and localization systems to balance system performance and affordability. The AMAS system has been demonstrated on six different families of tactical trucks.

3 Engineering the Automated Driving System

The systems engineering process is the starting point for conceptualizing, designing and testing automated driving. These days the research and engineering communities are transitioning from low levels of vehicle automation to higher levels while, at the same time, leading edge automated vehicle systems are transitioning from research to product development. Trucks fleets, both commercial and military, are likely to be early adopters of automated driving systems, driven by fuel costs, availability of drivers and an increasing emphasis on safety. The trucking market has been early adopter of key precursors to automated driving including systems like Roll/Yaw Electronic Stability Control, supplemental Electric Power Steering, Lane Departure Warning, Automatic Emergency Braking and Adaptive Cruise Control.

While research is often less structured in the early stages it becomes more structured in the later testing phases. Truck research is addressing higher levels of automation in systems like Traffic Jam Assist, Automatic Trailer Backing Assist, Freeway Pilot, and automated off-highway hauling and queue movement that are likely to be introduced the next 5 years. Engineers engaged in new product development follow a more highly structured systems engineering process with three phases: concept, development, and approval. Furthermore, developers will need to prove to the regulatory agencies (e.g., NHTSA, FMCSA, FHWA, Army Test and Evaluation Command, etc.) that these systems are safe and reliable before they are deployed.

The systems engineering process is presented in Fig. 1 along with methods for testing and evaluation depicted in red type below the associated activity box. In the concept phase system requirements guide the development of the initial proof of concept that is used to create a set of system specifications. Modeling and simulation are tools for testing the concepts for the hierarchy of systems, subsystems, software modules, and components.

It is beneficial to know up front how the system will be approved. One of the core challenges today in conceptualizing, designing, and testing the virtual driver is the lack of knowledge as to how the system will be approved and/or certified if that is required. Looking ahead to the future of approving automated vehicles for on road usage it is advantageous to prepare a plan for coordinating the system requirements and architectures with the approval processes and scope the test cases to take into account plans for later validation and verification.

Functional and non-functional requirements guide all phases of the process. The circles in the Fig. 1 indicate the iterative nature of the process. The concept phase generally leads off with an overall definition that includes a hazard analysis and risk assessment and then moves into development of a product and approval of a validated system. A formal validation and verification plan may include among other functional and nonfunctional requirements for truck automation:

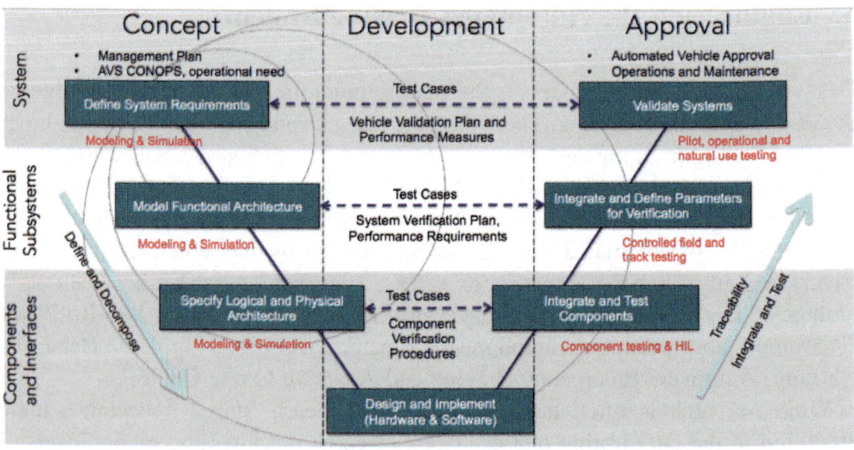

Fig. 1 Testing and evaluation in the model-based systems engineering process

- Functional correctness: Does the system deliver the specified functions (e.g., localization, navigation, leader/follower, obstacle avoidance, conforms to rules of the road, passenger riding comfort, vehicle safing, etc.) and maneuvers (e.g., following, passing, stopping, etc.,) in target environments including weather, road geometry, traffic signals and signs, road conditions, lighting, traffic, etc. (i.e., functionality, maneuverability)
- Fault management: What happens when a fault is detected and how is safety and dependability managed in that context? Does automated navigation require a real-time operating system and alternative path planning in cases of a severe fault? (i.e., safety, dependability, reliability, redundancy, etc.)
- Safety and dependability: Is the system sufficiently available, reliable, and maintainable to assure safety and avoid catastrophic consequences? How does the system manage faults and near to complete system shut down? (i.e., availability, maintainability, reliability, safety, etc.).
- Performability: Assuming there is no driver to take over, will the automated vehicle systems perform robustly under constraints like the presence of multiple faults? What is the vehicle safing strategy?
- Acceptable cost: Does the system address all the other requirements at a reasonable and acceptable cost in the context of expected production volumes and market demand, penetration, and competition? Does the system support well-defined interfaces that will promote component level competition on design and production? (i.e., extensibility, interoperability, etc.)

Reliability engineering for automotive virtual driver is challenging because of consumer cost sensitivity and most automotive equipment is used until the end of its life. Common practices such as triplex redundancy of critical components may not be affordable in automobiles. In most safety-critical tasks, preventive maintenance schedules call for replacing electronics before the end of their design life. In the automotive environment, many components are never replaced unless they fail. Furthermore, at highway speeds it would help to have cost effective sensors that could sense up to 300 m and comply with functional safety and all target environments. Many challenges still exist.

Once the requirements have been laid out the overall performance and safety of the product can be established through the verification and validation stages of the development process.

4 Validation and Verification

The purpose of most product testing is to ensure that the as-built systems and sub-systems meet the design requirements and that the product meets the requirements of the end customer. The definitions we are using for verification and validation originate from the Defense Modeling and Simulation Organization (DMSO) where verification is defined as "the process of determining if the model implementation and its associated data accurately represent the developer's conceptual description and specification." This requires some understanding the functional breakdown of the product. Whereas, validation is defined as "the process of determining the degree to which a model and its associated data provide an accurate representation of the real world from the perspective of the models intended use." In more common terms the main purpose of V&V is to answer key questions: (1) "are we building the right system?" (i.e., validation) and (2) "are we building the system right?", (i.e., verification).

Formal validation addresses whether the software or system does what the user really requires. Does it conform to the "customer" needs and the concept of operations?

Formal verification involves the specification of requirements and the design of a model or representation of the systems in a formal specification language that is semantically complete and allows for rigorous analysis. With the requirements expressed in a formal language the analysis can take on axiomatic or semantic forms. The axiomatic approach involves analytical reduction and mathematical analysis of the ability of the systems to achieve the specifications. The semantic approach involves model checking and uses exhaustive search through all possible program executions while looking for behaviors inconsistent with formally stated requirements. Both approaches are challenging when addressing automated driving in the context of functional safety.

The final evaluation and approval of these automated systems will address intended use of the automated system, use exceeding system limits, and use involving system faults. The evaluation depends on both the virtual driver's identification of critical situations and its ability to execute appropriate actions though a variety of maneuvers under a range of environmental conditions. The sheer magnitude of possible harsh environmental conditions, and possible critical situations poses additional serious challenges for design and acceptance of trustworthy virtual driving systems. An extremely improbable event, one that occurs once in 10 billion hour, would cause one failure in approximately 70 years in the US commercial air fleet. That same probability would cause a failure at least once a week in the US automotive fleet, due to the much higher number of vehicles. Even if the risk to a passenger might be the same in both cases, the public perception of risk could be much higher for road vehicles.

5 Trusting Automated Driving

Trust has a variety of meanings in systems engineering. In this context trust will assume a conventional meaning that is often used in trusted computing where the computer will behave consistently, in expected ways, and those behaviors will be enforced by the design of computer hardware and software. This is consistent with everyday usage where trust is learned from logical assertions and positive experiences over time. The problem for testing automated driving is to provide logic evidence of system performance, safety, and reliability over time under all reasonable conditions.

Compounding the problem of testing trust is that automated driving systems require advanced perception, navigation, and control systems that are distributed across many hardware and software components. Most automated driving systems are built with adaptive, non-deterministic "intelligent" algorithms to address a broad set of environmental conditions.

Complex software code often manifests isolated faults under rare scenarios that are difficult to find and expensive to produce through traditional V&V approaches described above. If the V&V strategy is to field test on roads then hundreds of millions of test-miles are unlikely to detect low-rate systematic defects in software that operates an entire fleet of vehicles. Furthermore, certain events pose challenges to the virtual driver as well as the human driver including abrupt maneuvers at high speeds. Control systems that are generally optimized for smooth performance at cruise may not work for abrupt maneuvers in emergency situations. Furthermore, special systems and controllers need to be designed and adopted to cope with flat tires or loss of power for braking or steering. Moreover, engine and transmission dynamics are difficult to model at slow speeds while icy roads and other low friction surfaces always difficult to handle.

More complex systems provide addition challenges, as traditional V&V methods are difficult to scale with increases in system complexity, environmental diversity, and breadth of usage scenarios. Increasing system complexity demands more elaborate testing and additional expense to guarantee the safety and reliability of these systems of systems. The complexity of the software itself is a major driver of system complexity, and which according to Wagner and Koopman [2], exhibits:

- Millions of lines of code (planning and control of a driverless vehicle has considerably more operations than throttle control)
- "Cyclomatic" complexity needed to implement driving behaviors,
- Multidimensional-dimensional interfaces to transmit rich perceptual data
- New algorithms involving real-time control, machine learning, and adaptation.

This suggests that analytical and simulation test scenarios, including maneuver features and environmental conditions, can be rated for their pervasiveness, abruptness, delta speed, and presence of challenges that will not only provide insight into the specific behavior being tested but possibly suggest outcomes when conditions are changed. For example, while there will be exceptions, successful performance of a maneuver at a high speed could suggest that the same maneuver could be performed successfully at reduced speeds. An approach work considering is to thoughtfully "push the envelope" in the simulated environment to enhance our understanding of performance limits.

6 Model-Based Systems Engineering

The design and evaluation of automated driving require advances in systems engineering to manage the system complexity and improve traceability of changes and impacts throughout the process. Modeling and simulation techniques are relevant to testing the performance of automated vehicles cooperating in road networks with traffic control, vehicle-to-vehicle coordination as in CACC, convoys, and platoons, individual automated vehicles on various road environments and responding to dynamic or stationery obstacles, automated vehicle components, subsystems, and software modules including hardware in the loop, networks of wire line and wireless channels including databus traffic at interfaces and V2V and V2I communication, and human behavior in the vehicles to the extent it is relevant in the higher levels of automation. Furthermore, these forms of simulation can be creatively integrated to address specific issues. In the context of vehicle automation simulation is a model-based approach for helping address a broad range of automated scenarios or identification of specific "worst case" scenarios. It can also be used for more in-depth study of specific areas, as well as testing new features prior to implementation. New systems dynamics models with feedback loops can help the Tester address complex systems relationships and feedback mechanisms. The Systems Modeling Language (SysML) is a general-purpose modeling language that supports the model-based approach including the specification, analysis, design, verification and validation of automated vehicle systems.

Model checking is a key feature in trusting automated real-time control systems [3]. Wagner and Koopman [2] predict that automated driving will rely heavily on inductive inference and complex software to operate safely and that traditional software safety techniques are not up to the task of analyzing and mitigating the risks they pose. Wagner argues software testing should focus more on negative test results that motivate ongoing, iterative software improvement [2]. The model-based approach opens the door for more creative and automated generation and testing of scenarios. An important rule for software engineering is that a defined output should be provided for any possible input. So instead of testing how the architecture handles normal, "clean" inputs, the model-based approach can test abnormal, "noisy" inputs crafted to stress the system and identify potential vulnerabilities including software bugs, flawed architecture design, communication failures, environmental conditions that exceed the systems design parameters, inconsistent internal states, and gaps in system testing.

TARDEC has used this approach to test their automated vehicle systems by feeding unexpected inputs into the sub-system to identify abnormal behaviors. This reveals problems that often go undetected using other forms of testing. Examples might include safety invariants such as speed limits or not moving while the emergency stop is engaged. The model can look at violations of these and other tests and flag whether the vehicle exceeds the speed limit or violates other safety rules or if the subsystem experiences faults the lead to system crashes. In general the Tester can develop a test specification based on the architecture including an interface definition and safety invariance that will be tested based on existing safety requirements. This information is used to generate a set of test cases during the test the system monitors output at run time for violations of the safety invariance.

7 Automated Driving Reference Architecture

The SAE Reference Architecture and Interfaces (RAI) Task Force is using Systems Modeling Language (SysML) general purpose modeling language to support specification, analysis design, verification and validation of the Automated Driver systems of systems. The SAE Reference Architecture uses nominal requirements, desirable properties, behavioral diagrams, threshold values for metrics, and structured diagrams to generate a SysML model that will aid in the analysis and requirement traceability required for verification and validation. This approach is being considered for model based identification and evaluation of real options to support strategic test planning. Model-based testing (MBT) relies on models of a system under test and/or models of its environment to derive test cases for the system [4].

A Reference Architecture helps facilitate a Model-Based Systems Engineering process in several ways: It provides a common language for systems design. It also provides best practices and design patterns that can aide in the development of safety and interoperability requirements for the virtual driver, particularly in early stage programs where many requirements are not well defined. It is a valuable tool for the systems engineering tool chest.

The SAE taskforce is continuing work started under the Interoperability Profile (IOP) activity initiated at U.S. Army Tank Automotive Research, Development and Engineering Center (TARDEC), Warren, Michigan. This taskforce is chartered to design high-level functional reference architecture for automated wheeled ground vehicles (e.g., military and civilian, trucks and passenger vehicles) covering use cases present in SAE Automation Levels (SAE J3016) 3 through 5 [5].

The RAI taskforce will identify possible standard work products and provide recommendations for extending existing vehicle messages sets (e.g. SAE AS-4 (JAUS), SAE J1939, SAE J2945, etc.) to encourage interoperability and reuse of automated vehicle of subsystems and components. Elements of existing standards like AUTomotive Open System ARchitecture (AUTOSAR), real-time operating system OSEK/VDX, Robot Operating System (ROS), LIN, CAN, FlexRay, and Ethernet are being leveraged to the extent that they applicable to fully automated vehicles.

Systems based on the reference architecture will need to address the hard-real-time requirements of vehicle control while intelligent monitoring of the driving environment on public and other drivable roads. This includes behaviors such as highway driving, obstacle avoidance, leader-follower, platooning, all the way up to full automation (i.e., SAE Level 5). The task force is following a process that includes the following: (1) review and mine patterns from state-of-the-art system designs, (2) analyze and decompose requirements, and (3) model and integrate a scalable, flexible, functional reference architecture the supports these requirements. The first phase will be documented in a whitepaper that describes the working reference architecture and its requirements, modules, interfaces, and messages. While the first phase focuses on prior art the second phases focuses on capturing specific requirements, such as those described in the introduction, and derived requirements for elements such as spatial coordinate frames, real-time control system hierarchies, data management and high-fidelity maps, communications analysis, trust and reliability, and possible performance parameters. This background and analysis is feeding into the third activity, the modeling of a scalable functional architecture using SysML. It is intended that this activity will continue to evolve as the technology and requirements evolve and as new best practices are adopted.

Figure 2 shows the functional layout of the working draft of the reference architecture supporting the entire dynamic driving task. RAI's current focus is on the development of SysML models for Level 4 for vehicles including cars and trucks. Figure 2 covers functional blocks of the architecture such as sensing, perception, navigation, active safety and driver assist, and vehicle controls. The details

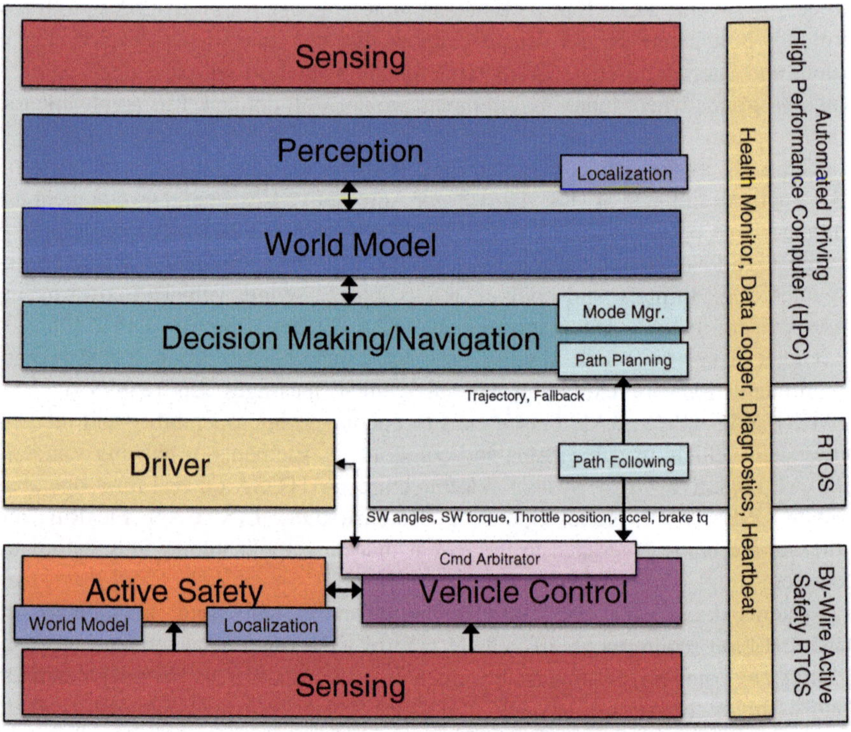

Fig. 2 On-road automated vehicle reference architecture scope

of diagram were selected due to their wide use across the published literature or as lessons learned from actual implementations. This diagram is provided more for discussion than for definition. Today's Active Safety and Driver Assist (orange in Fig. 2) provide not only provide a baseline for automated functionality they may also evolve independently from higher automation systems, and provide assistance to human and proving a secondary check on virtual drivers. It is assumed that in a physical implementation critical modules will have redundancies of some form, these have been omitted from Figs. 2 and 3 to reduce their complexity. The colors in Fig. 2 are coordinated with the layered view of the reference architecture in Fig. 3.

The ORAV reference architecture will provide a structural framework and common language that will assist in consistent modular implementation and possible standard interfaces. The concept is similar to how AUTOSAR provides Application Interfaces that can aide verification, system interoperability, and module reuse. Similarly solutions emerging from RAI are designed to support both module interoperability and subsystem functionality.

The critical analysis of the performance requirements of key automated vehicle functionality will drive certain design elements of the interfaces between various software and hardware modules. A particular emphasis is being placed on the interface between the virtual driver and the vehicles by-wire control system. In order to insure that these interfaces will support robust real world systems RAI has found it necessary to analyze the critical elements of vehicle safing behavior, motion controls, message latency, extensibilty/expandability, cybersecurity, fault detection and tolerance, and safety critical data bus traffic.

The ORAV reference architecture is a work in progress. While at this stage there is no plan to produce a standard RAI is intended to be an authoritative source of information that captures best practices that can guide and possibly constrain

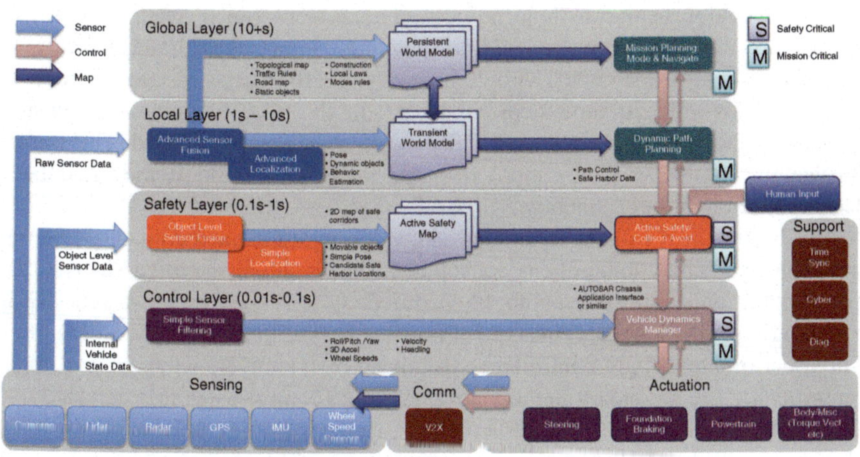

Fig. 3 On-road automated vehicle architecture layers

instantiations of solution architectures (architectures of systems being developed). An useful reference architecture will state its purpose and offer guidelines and logic for how requirements will be achieved through a description of reusable patterns of activity, functionality, information/data flows, and scenarios of sequenced responses to time events. It will provide guidance and structure at the logical level of (software) modules and messages that flow between the modules while taking into account the general use of the product and the related requirements, functions, and activities.

An important early focus is to evaluate the impact that system safety has on the design. The central approach to functional safety is to design the automated driving system so as to minimize unacceptable risk of physical injury or of damage to the health of people either directly or indirectly [6]. Timely driver takeover cannot be guaranteed in levels 4 and 5 automated vehicle systems. This means that these systems must be able to continue to operate safely, even under fault conditions, until they can be brought to a safe state. This requires identifying hazards and the required safety functions, assessing the risk reduction required by the safety function, and implementing appropriate risk mitigation techniques to ensure fault-tolerant performance in all conditions. Again, this is a challenging task.

The high degree of complexity required for vehicle automation combined with a high degree of safety-critical performance makes the verification a daunting task. The ORAV reference architecture leverages key techniques from NIST's RCS (Real-time Control System) framework [7, 8]. The RCS framework was also used to develop the 4D-RCS architecture for automated systems used by the DoD. Classical systems engineering manages verification by decomposing complex systems into subsystems, often along functional lines. The RCS framework seeks to achieve the balance between intelligent system performance and ease of verification by decomposing systems both by functionality and by time domain. A control system is decomposed into multiple layers based on a given functions temporal/spatial span-of-control requirements [7]. Each layer is allowed to plan forward and look behind a finite number of time steps (NIST recommends 10). Higher layer have time steps an order of magnitude longer than the layer below it. For example a layer may work with 20 time steps (± 10) of 100 ms each. The layer below it deals with 20 time steps of 10 ms. In this way continuous control is achieved while the complexity and state-space of each layer is bounded. This makes verification complex systems more tractable than they would if they had been decomposed along only functional lines.

This is combined with the functional safety concept of "mixed criticality". In this approach final safety authority is consolidated in key highly verified subsystems. This allows other subsystems to be certified at lower criticality levels while bounding system-level risk. This type of practice is used in air systems, where often a high-performance flight controller is shadowed by a simpler, safety-critical flight controller(s). The secondary flight controller(s) take control when it is detected that the primary flight controller has issued faulty, inappropriate or no commands at all. The secondary controller must maintain control until the primary system is recovered or the aircraft brought to a safe state (landed). While the complex, the primary flight controller is capable of high degree of performance (fuel economy, smoothness, etc.), the simpler design of the secondary flight controller allows it to be verified.

Figure 3 shows the same system as Fig. 2 decomposed into layers based time horizon and safety criticality. The proposed RAI architecture is broken into four notional layers, starting from the bottom:

1. Control Layer: The lowest/fastest layer deals with actuator control and vehicle dynamics management based mostly on internal state sensors (e.g., acceleration, gyro, wheel-speed, etc.). This would include traditional features like ABS, and ESC. This layer would need to be fault tolerant and hard-real-time.
2. Safety Layer: This layer the deals with simple maneuvers based on simple, robust external sensor data (such as radar object tracks). In manual vehicle operation is provides traditional ADAS and collision avoidance functionality. Under automated control it acts as the safety critical backup to the Local Layer. This layer will execute the Safe-Harbor maneuver in the event of the failure of higher level automation. An Active Safety Map (e.g. ADASIS) could be used to aid operation and Safe-Harbor functionality. This layer would need to be fault tolerant and hard real-time.
3. Local Layer: This layers deals with automated driving within sensor range, including fusing high-level sensor data, tracking other vehicles, projecting their behavior, populating the Dynamic Map, and planning vehicle maneuvers. It may also provide vehicle safing artifacts to the safety layer (such as suggested vehicle safing path).
4. Global Layer: This layer deals with beyond-sensor horizon planning, perceiving the static elements of the world around the vehicle, maintaining the long term static map, and planning the driving route (mission planning)

A rough analogy can be made with the human nervous system [9]. The Safety and Local layers are similar to the dorsal/ventral model of human perception. The dorsal visual system is a high-speed, highly deterministic pathway deals with core collision avoidance and precise tracking. The dorsal system can controls a person's movement without conscious input. This is in contrast the slower, less deterministic ventral visual system that deals with recognition, and identification and is a core input to conscience decision-making. The Global Layer and Control Layer are similar to prospective memory/planning and the cerebellum respectively.

8 Standards and Certification Testing

Testing will be needed for certification to specific standards developed for automated driving systems and to provide evidence for meeting future regulatory requirements. This may include voluntary standards, industry verification and validation processes, and government assessment processes. In the automotive industry this often involves laboratory testing of components combined with controlled tests on the test track.

In the United States, automotive safety certification is the responsibility of the OEM and their suppliers. Certification is based on compliance to engineering standards and recommended practices, as established by the organization (including SAE, IEEE, and TMC), and each vehicle/vehicle equipment manufacturer must test and certify that each motor vehicle and/or equipment item is in full compliance with the minimum performance requirements of all applicable Federal Motor Vehicle Safety Standards (FMVSS) set by NHTSA (e.g., Code of Federal Regulations, Title 49, Part 571). FMVSS are federal regulations specifying design, construction, performance, and durability requirements for motor vehicles and regulated automobile safety-related components, systems, and design features. Manufactures must confirm with the NHTSA that their products conform to the relevant standards through the process of self-certification. The NHTSA can inspect any product at any time to evaluate whether a vehicle or equipment item conforms to the performance requirements. For example, FMVSS 121 applies to trucks, buses and trailers equipped with air brake systems, but with some exceptions based on size, speed and weight, and FMVSS 105 applies to multi-purpose passenger vehicles, trucks and buses with a gross vehicle weight rating of 3.5 tons that are equipped with hydraulic or electric brake systems. The standards specify the test procedures that must be used for the purposes of self-certification, which include ASTM and SAE as well as procedures defined in the FMVSS document. The New Car Assessment Program (NCAP) also supports some voluntary third party testing and certification of automobile safety. Unlike the US, in Europe it is common practice for most testing to be done by third parties.

Automated driving is different from traditional active safety features because instead of aiding a driver, the system takes on the full driving task. Current standards, such as ISO-26262, assume that a human driver is in control and can mitigate faulty systems as long as these systems exhibit fail-safe behavior. Higher levels of automated driving must mitigate their own faults, maintaining fail-operational performance until the vehicle is brought to a safe state. At a conceptual level, tests for validation of automated driving Levels 4 and 5 is more analogous to human driver testing and may address, among other things:

- Basic maneuvering on surface streets, freeways, at intersections, in parking lots, etc.
- Maneuvering the vehicle to safety in case of a hazard or vehicle malfunction (i.e. minimal risk condition or safe state)
- Defensive driving and crash avoidance,
- Compliance with rules of the road, and
- The ability to recognize and handle complex, previously unseen scenarios.

Most of these fall under the category of performance testing with experimental controls providing an unbiased way of presenting the benefits or potential dangers of active safety systems on the test track. This approach requires the selection of specific geometric road configurations and accompanying traffic scenarios based on performance requirements. It also requires specification of driver behavior and vehicle maneuvers as well as the positions of target vehicles and other obstacles in the scenario.

One could speculate that it might be sufficient for safety certification to prove the automated driving system can detect faults and bring itself into a safe state under all specified situational and fault conditions. In the previous section we mention the simpler safety-critical controller in the Safety Layer assuming this role and serving as a check on the Local Layer trajectories and Global Layer plans. Certification of the system could focus on the performance of the Safety Layer and the Vehicle Safing software.

In addition to performance and behavioral certification standards there are likely be industry standards dealing with structures, devices, and interfaces. These may include elements described in the reference architecture such as interfaces between the virtual driver and the vehicle sub-system including the vehicle control systems, sensors, and communication devices. Other examples include standards for sensor interfaces, 3D prior localization maps, diagnostics, and event data recording. Because of the number of potential use cases and the complexity of the software, the regulating agencies need holistic automated driver testing strategies/methods. These may include model-based approaches supported by simulation or hardware in the loop (HIL) testing, as well as "edge case" or "worst case" philosophy in the selection of test scenarios.

9 Demonstration Pilots and Operational Testing

While low mileage pilots can demonstrate the feasibility of automated driving systems in controlled environments, human supervised high mileage operational testing offers opportunities to encounter novel situations in the on-road environment, to make design modifications based on road experiences, as well as to validate and increase trust for vehicle safety in realistic environments over time. The pilot projects of automated driving will generally involve operational testing once the concept is in place and the selected automated system is well down the path of development. In other words, a proof of concept has been completed and the safety requirements are in place along with the verified design and functional Human Machine Interface (HMI). In most cases, the HMI will been verified through a concept simulation and driving situations will have been tested in a driving simulator. Test track evaluation using professional test drivers will also been completed. Tests for navigation include blind path tracking tests, perception-assisted path tracking test, and perception planning test in a broad range of environments and alternative use conditions. Operational tests usually include experimental designs with specific validation and verification goals that focus on error detection and ensuring compliance in typical operating environments. It facilitates consideration of environmental factors that influence system behavior and allows feature interaction.

More advanced approaches to operational testing of automated vehicles could include human and virtual driver learning and continuous system improvements throughout the test process.

TARDEC's Joint Capability Technology Demonstration (JCTD) demonstrated a 10 truck automated convoy with six different vehicle types. It tested these systems' ability to avoid simulated obstacles and pedestrians, and to mitigate accidents commonly caused by driver inattention or distraction. The evaluation also included user (driver and remote operator) interfaces. Cyber security is a serious issue for the DoD, and an extensive "red-teaming" of these systems has been performed. This included vulnerability testing with physical access (i.e. laptop plugged into vehicle systems) as well as high-power jamming of radios, lidar and automotive radar systems.

While much of the DoD operational testing in the United States is conducted on based or military test areas some of the testing for connected vehicle applications in road traffic will be conducted on the interstate highway system. At the time of writing the states that allow on-road testing includes California, Florida, Nevada, District of Columbia, and Michigan. The proximity of Michigan's I-69 corridor to TARDEC makes if a convenient site for on-road testing the vehicle-to-vehicle (V2V) and vehicle-to-infrastructure (V2I) communication features of the AMAS concept.

A planned on-road test in 2016 will investigate 4 truck automated convoys along the Michigan I-69 freeway corridor between Port Huron and Flint where the vehicles will be traveling in close proximity to one another. This convoy will use V2V communications between the linked pair of trucks and is not dependent on V2V communications from non-convoy vehicles. The control system enables shorter distance and time gaps between vehicles in the convoy in a freeway environment with the potential to reduce fuel consumption and emissions, increase road throughput, and minimize driver fatigue associated with the long driving durations. The lead vehicle of this convoy will be manually driven. The trailing vehicles will follow the lead in sequence with distances of 20–30 m depending on the vehicle speed and breaking capabilities. While in this test, the drivers of the following vehicles will monitor the system. In actual use these drivers will have been situational awareness as they can focus their attention on the broader situational context. Drivers of the following vehicle must be in position to resume control with an increase lead-time if a takeover request from the system occurs. In this test the driver the vehicle should be capable of reacting to emergency braking maneuvers of their leading vehicle (Fig. 4).

While operational testing may be required to increase consumer and public trust for automated trucking it may not be as important an element in acceptance as durability and reliability testing. It is widely accepted in the automated vehicle field that these vehicles will need an abundance of time and miles for simple durability and reliability testing. It may be that the real value of operational testing is the ability to observe the vehicles behaviors under a wide variety of situations. These observations would be used in the development of better tests and more effective behaviors. The combined value of testing and learning over many miles and long durations of time is the identifications of important edge cases while increasing trust in these systems as improvements are made.

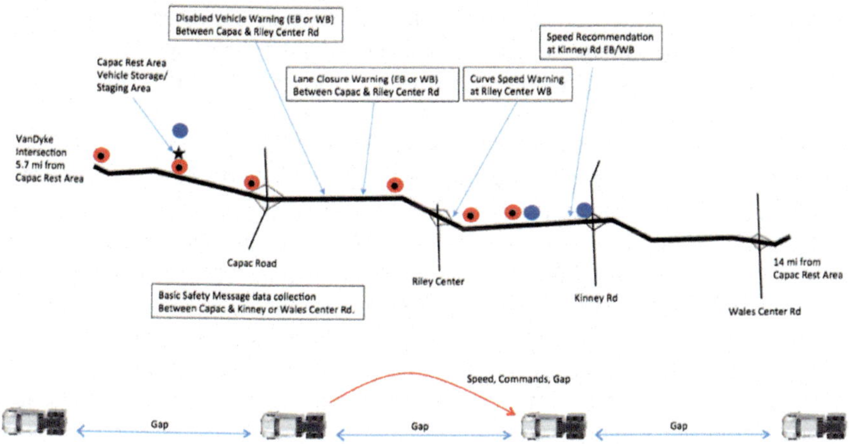

Fig. 4 Operational test for Michigan truck caravan

10 Conclusion

One of the themes at the Automated Vehicle Symposium was the identification of new approaches for testing and evaluating in support of efficient engineering, certification, and consumer acceptance of automated driving systems. This chapter details and expands upon this theme from the perspective of the military truck panel in the breakout session on Truck Automation. Although demonstration and testing of platooning was discussed at length in other panels other of this session the primary focus on this panel was unmanned leader and follower behavior of trucks in military convoys. We address the testing and trusting of the virtual driver, or automated driving system, at the higher levels of automation where it is presumed a human driver is not available as a backup to the automated driving system.

While the design and operation of self driving trucks poses research and engineering challenges the parallel development of new methods for testing and certifying the virtual driver may be the most difficult obstacle in the path leading the way to public acceptance and trust of truck automation. Although current modeling, simulation, testing and evaluation methods have been effective for millions of commercial and military trucks in service, they are much less effective for the evaluation of learning and non-deterministic systems are being developed to support higher levels of autonomy. New approaches are needed to address the complexity and diversity of systems and operating environments that automated trucks will be used in.

The professional community is working on developing a methods and procedures for testing and evaluating the automated vehicles that draws on historical methods of modeling and empirical testing while investigating new adjustments and strategies for testing the performance and safety of the virtual driver. There is a need for standard and accepted test procedures. While many good tools are available for modeling, testing, and simulation; however these can find problems but can't prove the goodness of an automated driving system. We conclude with several ideas that may help along out path to a new method for verification and validation of automated vehicles.

First, it is highly desirable to have a standard reference architecture that clearly defines and delineates the roles and functions of the virtual driver and the automated vehicle, as well as a standard interface between these entities. This would greatly facilitate the development of standards that could be used to validate and certify the higher levels of automated driving for use around human beings. In addition, this approach will enable continuous improvement of these systems, with OEMs, vendors, and regulators being able to focus on a common system, rather than diffuse efforts on multiple approaches. Furthermore, functional safety standards like ISO 26262 and its associated ASILs need to be extended and upgraded to take into account automated driving systems.

Second, system models must be developed to describe complete use cases and capture the function and structure of the system. These use cases can then be used to generate the proper test scenarios for V&V testing. The complexity of the driving environment makes objectively measuring risk difficult, and compounds the ever-increasing cost of redesign due to errors found in late developmental and operational testing. The diversity of the driving tasks makes it impractical to test under all conditions. Novel methods and approaches are needed to address this.

Third, high fidelity modeling and simulation must be heavily used to pre-test system behaviors, with specifically selected physical testing performed to validate the simulated responses. Forcing failures parallel simulations to expose issues. Better tools needed for massive (greater than terabyte) data collection, data mining and scenario recreation. Innovations in "worst cases" approaches to simulation and modeling are needed.

Finally, safe and correct operation of automated driving must be repeatedly demonstrated with real vehicles and broadly advertised to build trust and acceptance with the public. Extensive testing is required but not sufficient for dynamic real-time learning and adaptive automation systems.

References

1. Haddon W Jr (1980) Advances in the epidemiology of injuries as a basis for public policy. Public Health Reports 95(5):411
2. Wagner M, Koopman P (2015) A philosophy for developing trust in self-driving cars. In: Meyer G, Beiker S (eds) Road Vehicle Automation 2, Springer International Publishing, pp 163–171
3. Goldman RP, Musliner DJ, Pelican MJ (2000) Using model checking to plan hard real-time controllers. AIPS Workshop on Model-Theoretic Approaches to Planning
4. Utting M, Pretschner A, Legeard B (2012) A taxonomy of model-based testing approaches. Softw Test Verif Reliab 22(5):297–312

5. SAE International, J3016—taxonomy and definitions for terms related to on-road autonomous vehicles. http://www.sae.org/works/documentHome.do?docID=J3016&inputPage=wIpSdOcDeTaIlS&comtID=TEVAVS
6. ISO 26262–3:2011(en) Road vehicles – Functional safety – Part 3: Concept phase, International Standardization Organization
7. Albus JS (2002) 4D/RCS – a reference model architecture for intelligent unmanned ground vehicles. In: AeroSense. International Society for Optics and Photonics, pp 303–310
8. Albus J, Huang HM, Lacaze A, Schneier M, Juberts M, Scott H, Murphy K (2002). 4D/RCS: A reference model architecture for unmanned vehicle systems version 2.0. National Institute of Standards and Technology (NIST). US Department of Commerce, Gaithersburg, Maryland 20899
9. Albus J (2008) Toward a computational theory of mind. J Mind Theory 1(1):1–38

Automated Vehicles: Take-Over Request and System Prompt Evaluation

Myra Blanco, Jon Atwood, Holland M. Vasquez, Tammy E. Trimble, Vikki L. Fitchett, Joshua Radlbeck, Gregory M. Fitch and Sheldon M. Russell

Abstract As automated driving technology advances, the driver's role continues to shift from active vehicle control to passive monitoring of the automated driving system and environment. This study comprised three experiments on controlled test tracks in mixed traffic that investigated driver interactions with prototype Level 2 and Level 3 partially automated driving systems. The study investigated which human-machine interface (HMI) characteristics are most effective at issuing a Take-Over Request (TOR) during the operation of a Level 2 automated driving system, how to prompt drivers to attend to the road when distracted during the operation of a Level 2 automated driving system (and whether these prompts are effective over time), and which HMI characteristics are most effective at issuing a TOR during the operation of a Level 3 automated driving system. In addition, participants' trust in the automated driving system they experienced was gauged through multiple Likert-type surveys and an after-experience interview.

M. Blanco (✉) · J. Atwood · H.M. Vasquez · T.E. Trimble · V.L. Fitchett · J. Radlbeck · G.M. Fitch · S.M. Russell
Virginia Tech Transportation Institute, 3500 Transportation Research Plaza, Blacksburg, VA 24061, USA
e-mail: mblanco@vtti.vt.edu

J. Atwood
e-mail: jatwood@vtti.vt.edu

H.M. Vasquez
e-mail: hvasquez@vtti.vt.edu

T.E. Trimble
e-mail: ttrimble@vtti.vt.edu

V.L. Fitchett
e-mail: vfitchett@vtti.vt.edu

J. Radlbeck
e-mail: jradlbeck@vtti.vt.edu

G.M. Fitch
e-mail: gfitch@vtti.vt.edu

S.M. Russell
e-mail: srussell@vtti.vt.edu

G. Meyer and S. Beiker (eds.), *Road Vehicle Automation 3*,
Lecture Notes in Mobility, DOI 10.1007/978-3-319-40503-2_9

Keywords Automated driving technology · Human factors · Human-machine interface · Levels of vehicle automation · Self-driving vehicles · Vehicle automation

1 Introduction

Automotive manufacturers are currently deploying automated driving systems (ADS) as standard and optional features on vehicles at various price points. ADSs are those in which at least one aspect of a safety-critical control function—for example, steering, throttle, braking—is sustained without driver input. The National Highway Traffic Safety Administration (NHTSA) has defined five levels of automation which serve to describe a continuum of vehicle control. The NHTSA levels are shown in the Fig. 1.

2 Objectives

Researchers at the Virginia Tech Transportation Institute (VTTI), in collaboration with General Motors and Google, studied how operators interact with Level 2 and Level 3 ADSs, and how this interaction is affected by the human-machine interface (HMI).

This project involved the use of three different prototype ADSs; two of which were Level 2 systems, and one which simulated a Level 3 system. Three experiments were conducted on test tracks simulating highway driving. Experiments 1 and 2 were conducted at General Motors' (GM) Milford Proving Ground (MPG) in Milford, Michigan. This 4.5-mile banked circle track comprises five lanes, each with a designated speed range. Traffic from other studies was present at the time this

Fig. 1 NHTSA's five levels of automation

study was being run. Experiment 3 was conducted on the 2.2-mile Virginia Smart Road in Blacksburg, Virginia. Confederate vehicles were present on the track at the same time.

The methods implemented used real vehicles in mixed-traffic conditions to ensure that the consequences of a potential crash were present at all times. This served as an incentive for operators to behave as similarly as possible as to how naïve drivers would behave when using ADSs in real-world conditions.

3 Testing

3.1 *Experiment 1: Alerting Operators to Regain Control of a Level 2 Automated Vehicle*

3.1.1 Conduct of Experiment

This experiment investigated which characteristics are most effective at issuing TORs of Level 2 automated vehicles. Twenty-five participants drove a 2009 Chevrolet Malibu with a prototype Level 2 system on the MPG circle track for one 90-min session. Participants were provided with an Asus Nexus 7 tablet computer and the in-vehicle experimenter provided a series of navigation, email, and Web-browsing tasks to be completed while the vehicle was in Level 2 automation.

At approximately 5-min intervals (times ranged from 2 to 8 min), participants were provided with unimodal (visual only) or multimodal (visual plus haptic seat vibration) alerts instructing them to take control of the vehicle. Participants were told that they could activate or reactivate the automation whenever they were comfortable doing so. Alerts were of three types: Cautionary, Imminent, and Staged.

Cautionary alerts were 30 s in length and comprised a yellow LED and a "Take Steering" message. In addition to those visual signs, multimodal alerts incorporated a haptic seat alert lasting approximately 800 ms. If the participant did not regain control of the vehicle within 30 s of the onset of the alert, the alert would turn off. Imminent alerts were also 30 s in length and consisted of a red LED and a "Take Steering" message. As with the Cautionary alerts, the multimodal alerts incorporated a haptic seat alert lasting approximately 800 ms; the alert turned off if the participant did not regain control of the vehicle within 30 s.

The Staged alert was a combination of the Cautionary and Imminent alerts. The cautionary alert phase of the Staged alert lasted 10 s and included the yellow LED and the "Take Steering" message (the multimodal version included an 800-ms haptic portion). If the participant did not regain control within 10 s of the onset of the alert, the Imminent alert phase commenced. This consisted of the red LED and the "Take Steering" message; the multimodal Staged alert also included the haptic seat vibration.

Table 1 Summary of significant findings for Experiment 1

Variable	Alert modality	Alert type	Remarks	Take away: range mean values
Time to react	Sig		• Faster reaction when alerts were presented in a multimodal fashion	0.6 s (SM)–3.7 s (CU)
Time to regain control	Sig	Sig	• Faster time to regain control when alerts were presented in a multimodal fashion • Faster time to regain control with Imminent alerts	1.1 s (CM)–6.3 s (CU)
Time to activate automation	Sig	Sig	• Slower time to activate automation after Imminent unimodal alerts compared to Imminent multimodal alerts and Staged unimodal alerts	3.4 s (IM)–4.0 s (IU)
Time to release control of steering	Sig	Sig	• Faster time to release control of steering after Imminent multimodal alerts compared to cautionary and Staged alerts	2.4 s (IM)–4.6 s (SM)
Time to resume non-driving task		Sig	• Faster time to resume non-driving task after experiencing a Staged alert compared to an Imminent alert	2.3 s (CU)–3.1 s (CM)

S Staged, *C* Cautionary, *I* Imminent, *M* Multimodal, *U* Unimodal

Participants received 18 alerts: six each for Cautionary, Imminent, and Staged (three unimodal and three multimodal for each of these alert types). After experiencing the 18 alerts, an experimenter-injected lane drift was triggered. All 25 participants gained control of the vehicle prior to entering the adjacent left lane.

3.1.2 Results

Results of this experiment indicated that participants reacted to and regained control of the vehicle faster after the visual + haptic (multimodal) alert than they did after the visual-only (unimodal) alert. In addition, participants took less time to reactivate the automation after a multimodal Imminent alert than they did after the unimodal Imminent alert. The experiment's results are summarized in the Table 1. The alert modalities were unimodal and multimodal, and the alert types were Cautionary, Imminent, and Staged.

3.2 Experiment 2: System Prompt Effectiveness Over Time

3.2.1 Conduct of Experiment

Experiment 2 investigated how to prompt operators to monitor the road when operating a Level 2 automated vehicle as well as the effectiveness of prompts over time. Fifty-six participants drove a 2010 Cadillac SRX equipped with a prototype Level 2 automated system on the MPG circle track for three 60-min sessions. Participants were instructed to use the provided Asus Nexus 7 table computer to complete specific tasks at their leisure while the vehicle was in Level 2 automation. The non-driving tasks comprised 30 each in these categories: navigational, email, and Web browsing, and they were similar in terms of the visual/manual demand required.

Participants were assigned one of these prompt conditions: 2-s, 7-s, or No Prompts. When the participant's attention state was off-road, the driver monitoring system provided alerts based on the assigned prompt condition (e.g., for those assigned to the 2-s prompt condition, the prompts began after the participant's attention state was off-road for 2 s). Participants assigned to the No Prompts condition did not receive any prompts.

Alerts were issued in stages. Stage 1 was activated once the participant's attention was off-road for the threshold set by the prompt condition and was characterized by a lower-urgency visual alert. If the participant did not change his/her attention state to on-road within 5 s, the second stage began. Stage 2 was characterized by a higher-urgency multi-modal (visual and haptic) alert. If the participant did not change his/her attention state to on-road within 5 s of the onset of Stage 2, the third stage began. Stage 3 was characterized by a higher-urgency multi-modal (visual, haptic, and auditory) alert. The Stage 3 alert continued until the participant took control of steering.

Each participant experienced two experimenter-injected lane drifts that were initiated when the participant was involved in a non-driving task. One 60-min session included a lane drift that was accompanied by an alert (visual plus haptic), and one session included a lane drift that did not have an alert. One of the 60-min sessions had no lane drift. The lane drifts with the alerts represented situations in which the system warns the driver of a lane-keeping performance issue; the lane drifts without alerts represented a lane keeping performance issue along with a prompt system failure.

3.2.2 Results

When performing non-driving-related tasks, 47 % of the participants that received a lane drift without an alert had to be assisted by the experimenter to take over the vehicle control because the participant didn't respond to the lane drift in a timely manner (i.e., they had performed a lane incursion into the adjacent lane of at least half of the adjacent lane). When participants experienced the visual plus haptic alert along with the lane drift, they reacted 66 % percent faster than when they experienced the lane drift without the alert.

The experiment's results are summarized in the Table 2. There was no evidence of an effect of event type or prompt condition on time to release control or time to resume a non-driving task.

Table 2 Summary of significant findings for Experiment 2

Variable	Event type	Prompt condition	Performance over time	Remarks	Take away: range mean values (LD)
Time to react	Sig			• Faster reaction time to lane drift when presented with an alert	1.0 s (7A)–4.3 s (NPS)
Time to regain control	Sig			• More likely to regain control before passing into the next lane with an alert	2.0 s (2A)–5.0 s (NPS)
Time to activate automation	Sig		Sig	• In early driving sessions, experiencing an alert along with a lane drift may result in longer time to activate the automation; however, this relationship may not last as time goes on	3.8 s (2S)–8.2 s (7A)
Monitoring rate		Sig	Sig	• If participants were not engaged in the roadway at some point in time, prompts significantly increased participants' monitoring rate	19.3 % (NP)–31.5 % (2), compared after prompt
Non-driving-related glances	Sig	Sig	Sig	• The amount of non-driving-related glances decreased with time • Experiencing prompts increased the amount of glances participants made	2.5 g (NP)–3.5 g (2), compared after prompt

S Silent Lane Drift, *A* Lane Drift with Alert, 2 = 2-s Prompt, · 7 = 7-s Prompt, *NP* No Prompt

3.3 *Experiment 3: Human-Automation System Performance Over Time*

3.3.1 Conduct of Experiment

Experiment 3 examined how operators in a mixed-traffic simulated highway driving condition recognize and respond to hazards and how different visual allocation strategies influence their response. This experiment studied how operators respond to alerts and driving scenarios, including how and when they take control from the automation and release control back to the automation. Twenty-five participants drove a 2012 Lexus with a prototype system that simulated Level 3 capabilities for three 30-min sessions on the Virginia Smart Road in Blacksburg, Virginia.

Prior to entering the test track, participants viewed a 10-min video which summarized the vehicle's features, operation, and automated components. Once in the vehicle, participants were provided with an Asus Nexus 7 table computer that was loaded with games, movies, and Internet access. No specific tasks were assigned to participants but they were instructed that they could use it and/or their cellphone as much as they deemed appropriate, but only when the vehicle was in Level 3 automation.

At any given point in time on the test track, there were one or two confederate vehicles in close proximity to the subject vehicle; these vehicles entered and exited the participants' driving path at various intersections and points. The confederate vehicles presented in various configurations (lead, lead and side, side) that were unpredictable to the participant.

Participants experienced three different types of alerts, one in each 30-min session: Staged, Imminent-External Threat, and Imminent-No External Threat. The Staged alert for this experiment comprised 4 phases: an informational message phase (20 s), two cautionary alert phases (10 s each), and an imminent alert phase (10 s). The visual and auditory messages delivered during the Staged Alert became increasingly more urgent with each phase. All participants reacted to the alert by taking control of the vehicle prior to the beginning of the third phase; thus, prior to the imminent alert phase.

The Imminent-External Threat alert was triggered by the in-vehicle experimenter to alert the participant to an obstacle in the roadway ahead (a box that had been dropped surreptitiously by a confederate vehicle). Another confederate vehicle that was in front of the subject vehicle would swerve out of the lane to suddenly reveal the box at approximately 10 s time to collision (TTC). The participant could take control of the vehicle by braking, accelerating, turning the steering wheel, or pressing the automated system's off button.

The Imminent-No External Threat alert was triggered by the in-vehicle experimenter to simulate a warning due to a system problem. No details were provided to the participant about the reason for the alert. The Imminent-External Threat and Imminent-No External Threat alerts were both 10 s in length.

3.3.2 Results

The experiment's results are summarized in the Table 3. There was no evidence of an effect of alert type or driving session on time to activate automation or time to resume a non-driving task. Performance over time was a variable of interest; however, the experiment revealed no evidence of a significant change in performance over time.

Table 3 Summary of significant findings for Experiment 3

Variable	Alert type	Remarks	Take away: mean values
Time to react	Sig	• Those experiencing a Staged alert may take longer to react than they would to Imminent alerts, but this may only be if they are looking forward at the time of the alert	0.7 s (INT)–1.2 s (S)
Time to regain control	Sig	• Those experiencing a Staged alert may take longer to regain control of the vehicle than they do after Imminent alerts	2.1 s (IET)–17.0 s (S)
Time to release control of steering	Sig	• Those experiencing a Staged alert may take less time to release control of the steering compared to when they experience an Imminent alert combined with an external threat, but not necessarily longer than an Imminent alert when an external threat is not present	1.4 s (S)–2.7 s (IET)
Non-driving-related glances	Sig	• The rate of off-road glances significantly decreased immediately after a Staged alert and increased immediately after an Imminent alert with an external threat	0.1/s (S)–0.4/s (INT), compared after alert

S Staged, *INT* Imminent No External Threat, *IET* Imminent External Threat

4 Conclusions

The purpose of this study was to investigate user interactions with Level 2 and Level 3 ADSs. The study focused on how operators transition between automated and non-automated vehicle operation and how this interaction is affected by the HMI. The study shows that appropriate interactions with Level 2 and Level 3 ADSs are possible providing effective HMIs are utilized and the findings suggest that the most effective hand-off strategies were those that incorporated nonvisual components.

In addition, participants' trust in the systems they experienced was gauged through multiple Likert-type surveys and an after-experiment interview. The feedback received from participants suggests a high level of trust in automation and alternatives to present information to the operators in these vehicles that could effectively assist them to react and regain control when needed.

This study's results may provide designers and practitioners with methods for alerting drivers as well as the time that take control decisions might take for a majority of users (e.g., 90th percentile) if the priorities are potentially shifted in favor of other convenience tasks such as email or Web browsing. Effective HMIs will need to balance conspicuity, urgency, and annoyance. Future research could provide a more complete understanding of reaction time patterns for operators of vehicles with Level 2 and Level 3 ADSs. The final report for this study [1] can be accessed from the NHTSA website.

Acknowledgments This study was funded by the National Highway Traffic Safety Administration and the Intelligent Transportation Systems Joint Program Office. General Motors and Google served as partners on this project.

Reference

1. Blanco M, Atwood J, Vasquez HM, Trimble TE, Fitchett VL, Radlbeck J, Morgan JF Human factors evaluation of level 2 and level 3 automated driving concepts. Report No. DOT HS 812 182. Washington, DC, National Highway Traffic Safety Administration

Motion Sickness in Automated Vehicles: The Elephant in the Room

Cyriel Diels, Jelte E. Bos, Katharina Hottelart and Patrice Reilhac

Abstract Automation disuse and associated loss of automation benefits may occur if users of automated vehicles experience motion sickness. Compared to conventional vehicles, motion sickness will be of greater concern due to the absence of vehicle control and the anticipated engagement in non-driving tasks. Furthermore, future users are expected to be less tolerant to the occurrence of motion sickness in automated vehicles compared to other means of transport. The risk of motion sickness may be manageable if we understand underlying causes and design our vehicles and driver-vehicle interactions appropriately. Guided by three fundamental principles, an initial set of design considerations are provided reflecting the incorporation of basic perceptual mechanisms.

Keywords Vehicle automation · Design · Displays · Motion sickness · Carsickness · Sensory conflict · Anticipation

C. Diels (✉)
Centre for Mobility and Transport, Coventry University,
Priory Street, Coventry CV1 5FB, UK
e-mail: cyriel.diels@coventry.ac.uk

J.E. Bos
TNO Perceptual and Cognitive Systems, Soesterberg and Behavioural and Movement Sciences, Vrije Universiteit Amsterdam, C/O Kampweg 5, 3769 DE Soesterberg, The Netherlands
e-mail: jelte.bos@tno.nl

K. Hottelart · P. Reilhac
Valeo Schalter und Sensoren GmbH, Laiernstr. 12, 74321 Bietigheim-Bissingen, Germany
e-mail: katharina.hottelart@valeo.com

P. Reilhac
e-mail: patrice.reilhac@valeo.com

G. Meyer and S. Beiker (eds.), *Road Vehicle Automation 3*,
Lecture Notes in Mobility, DOI 10.1007/978-3-319-40503-2_10

1 Introduction

During his campaign in Africa, Napoleon quickly became to realize the strengths of camels as a new mode of transport and subsequently instated his dromedary regiment. What he did not quite foresee, however, was that the gait of the camel made his soldiers suffer from motion sickness. Camels clearly weren't for everyone with some generals refusing to use them for their troops.

Motion sickness may also prove to be a hindrance to the successful introduction of vehicle automation. The fundamental feature of handing over vehicle control in combination with the anticipated engagement in non-driving tasks will increase the likelihood that occupants experience motion sickness [1, 2]. In turn, this may result in *automation disuse* and consequently limit the ultimate socioeconomic and environmental benefits this technology has to offer.

This is not to say that motion sickness is a showstopper. The risk of motion sickness can be managed if we understand its underlying causes and design our vehicles and driver-vehicle interactions appropriately. Whereas this may seem obvious, a review of the automated vehicle concepts recently being put forward by design consultancies, suppliers, and OEMs, suggests that the risk of motion sickness has not been considered in the design process, a notable exception being the Valeo Mobius™ system [3].

The crucial starting point is the realization that motion sickness is a natural response to an unnatural motion environment. Man was designed to travel by foot. For reasons not quite understood, we don't like being exposed to motion environments which violate our finely calibrated relationship between the motion sensed by our eyes, organs of balance, proprioception, and ultimately our brain. Whereas the final manifestation of motion sickness is vomiting, it is typically preceded by signs and symptoms such as (cold) sweating, pallor, flatulence, burping, salivation, and apathy, after which nausea and retching may occur [4]. To this date, the exact evolutionary advantages of these responses remain a mystery although they most certainly motivate the sufferer from seeking a less provocative environment in the hope of a speedy recovery. Nonetheless, whereas we may not quite understand the *why*, we know fairly well which conditions lead to motion sickness and seem to understand the *how*. As with many other human physiological regulatory systems, the control of body motion is at issue, and an assumed error signal correlates quite well with observed sickness. On this basis we are able to provide guidelines for the design of automated vehicles.

Before discussing these in detail, we would like to spell out why automated vehicles are a special case. One could argue that automated vehicles are not different from other modes of transport. After all, what is the difference between being a train or car passenger and driving in automated mode?

First, unlike most car journeys, a large proportion of journeys on public transport tend to consist of long distance and uninterrupted routes at largely constant speeds. With the exception of perhaps some remote, rural areas, the daily reality is that our traffic on the roads can perhaps best be characterized summarized as stop-start. As explained in more detail below, it is this varying velocity motion profile that is particularly conducive to motion sickness. It also goes some way in explaining why even today we see very few car passengers reading books and instead prefer to stare at the vehicles in front.

Secondly, and most importantly, the ability to free up our time to engage in more useful or enjoyable activities is more pertinent to vehicle automation. Whereas we may elect not to use public transport or accept that we won't be able to use our laptop sitting in the passenger seat as it makes us feel queasy, the proposition of vehicle automation differs. The benefits of automation may not be perceived significant unless we can actually engage in other activities. In fact, the inability to do so, and in the worst case, having to constantly monitor the system and environment, may well be perceived as less comfortable and acceptable than manual driving.

Finally, it can be predicted that even if people do not experience full blown symptoms of motion sickness, mild and subtle symptoms may already negatively affect the user experience. As a consequence, people may be unable or unwilling to take advantage of the new scenarios that vehicle automation affords.

It is for the above reasons that we believe that motion sickness in automated vehicles should be considered a special case. We propose that the development of measures to minimize the severity of motion sickness, or avoiding its occurrence altogether, ought to become an important line of automotive research. Moreover, there is a matter of urgency in that the issue will be especially relevant during the introductory period in which the general public may be hypercritical with the least publically known failure easily leading to unwanted delays.

2 Drivers of Motion Sickness

In the below, we set out the conditions that are conducive to motion sickness in order to then provide guidelines for the design of automated vehicles and driver-vehicle interactions.

2.1 Vehicle Dynamics

Motion sickness typically occurs when we are exposed to motion that, from an evolutionary perspective, we are not used to, such as low frequency oscillating motion [5]. Whereas sea and airsickness are mainly caused by slowly oscillating vertical motion, carsickness is largely associated with horizontal accelerations caused by accelerating, braking, and cornering [6]. An aggressive driving style involving plenty of accelerating and braking is therefore more likely to result in carsickness. The implication for automation is that the vehicle dynamics in terms of acceleration/deceleration may have to be restricted to ensure a sufficient level of occupant comfort. As a general rule, motion profiles in the region of 0.16 Hz should be avoided [7]. It is noteworthy that this may compromise the expected benefit of automation on network capacity. For example, LeVine et al. [8] have shown on the basis of microsimulations that limiting vehicle dynamics will reduce signalized intersections' vehicle-processing capacity and increase delays.

2.2 Anticipation of Vehicle Dynamics

The motion profile becomes even more critical in automated vehicles due to the fact that the driver hands over longitudinal and/or lateral control to the vehicle. This transfer, or loss of control, lies at the heart of vehicle automation per se and in effect renders the driver a passenger. Unfortunately, we already know that drivers of cars, pilots of aircraft, or Virtual Reality users in control of their own movements usually suffer much less from motion sickness despite the fact that they experience the same motion as their passengers [9]. Thus, the mere fact of not being in control of the vehicle dynamics increases the likelihood that people will experience motion sickness.

Anticipation plays a key role in explaining this phenomenon. The difference between a driver and passenger can be understood by assuming our central nervous system not only reckons sensed motion, but also makes a prediction about self-motion based on previous experiences [7]. A discrepancy or conflict between integrated sensory afferents, and a prediction thereof by a so called internal model or neural store, is assumed responsible for generating motion sickness [4, 7]. If the driver of a car is familiar with the transfer from pedals and steering wheel input to the actual motion of the car, he or she can make a more accurate prediction, i.e., anticipate motorically about future motion, thus minimizing the sickening conflict. Although a forward looking passenger can see a curve ahead, it is only the driver who knows whether this curve will be taken wide or sharp, thus having optimal information about self-motion, resulting in the smallest possible and typically sub threshold conflict. Braking and accelerating will likewise cause a difference in conflict and hence a difference in sickness.

What is of particular relevance for the avoidance of motion sickness in future automated vehicles is that this anticipatory mechanism is not only at play when individuals are able to motorically anticipate incoming sensory cues, but also on the basis of visual information alone. A clear view of the road ahead will allow for the prediction of the future motion path at least to some degree and is therefore beneficial in reducing sensory conflict.

The importance of anticipatory visual information in motion sickness is suggested by the anecdotal evidence that backward looking passengers suffer more from car sickness than forward looking passengers, the former only seeing the trajectory that has been followed, the latter seeing the trajectory that will be followed. In addition, we also know that rear seat passengers are particularly prone to car sickness under conditions where external visual views are limited [10].

Recently, the effectiveness of visual information in reducing motion sickness was strikingly demonstrated in a flight simulator study [11]. No less than a fourfold reduction in motion sickness was observed when a visual, roller coaster-like track to be travelled was presented. Although it has yet to be determined whether this approach could be successfully applied within an automotive context, in theory, we would predict a similar approach to reduce or prevent motion sickness in automated vehicles. Possible design solutions may include augmenting the natural scene with a future motion trajectory via contact-analogue Head-Up Displays (HUD) for example. Incidentally, such an approach was recently adopted by Weißgerber et al. [12] albeit with a different goal. In their study, providing the driver with an augmented view of the road ahead indicating, amongst other things, the vehicle's future trajectory improved driver's ability to create a correct mental model of the driving situation and automated vehicle system.

From the above, it becomes apparent that a person's ability to anticipate the future motion trajectory is a major factor in the development of motion sickness. When traveling in automated mode, the absence of vehicle control, facing away from the direction of travel or even traveling backwards, or not having a clear view of the road ahead due to it being obscured by displays or internal structures otherwise, will all increase the likelihood of occupants experiencing motion sickness and should therefore be avoided.

2.3 *Conflicting Motion Cues*

Vehicle automation will ultimately enable the driver to engage in non-driving activities. It is likely that popular activities will include reading, responding to emails, or engaging otherwise with nomadic or integrated infotainment systems such as in-vehicle displays, laptops, gaming consoles, or tablets [13]. From the perspective of the user, increased comfort and the ability to spend the driving time more productive or enjoyable are arguably the main advantages of automation.

As alluded to already, engagement in such activities can be expected to lead to an increase in carsickness. Similar to reading a map or book whilst driving, the (static or dynamic) motion shown on displays may not correspond to the motion of the vehicle which, in turn, will increase the likelihood of motion sickness. The essential point here is that our central nervous system integrates visual and vestibular signals (i.e., originating from the organs of balance within the inner ear) normally caused by congruent motion inputs as expected. Watching a scene showing different motion than that felt by our organs of balance is not what we expect and this inter-modality sensory conflict has long been known to lead to motion sickness [4].

With regard to non-driving tasks, Cowings et al. [14] reported a negative impact on crew performance and health when subjects attended to displays while the vehicle was moving. Similarly, in a study by Kato and Kitazaki [15] participants were driven around whilst sitting in the backseat either watching the road ahead, or a rear-seat

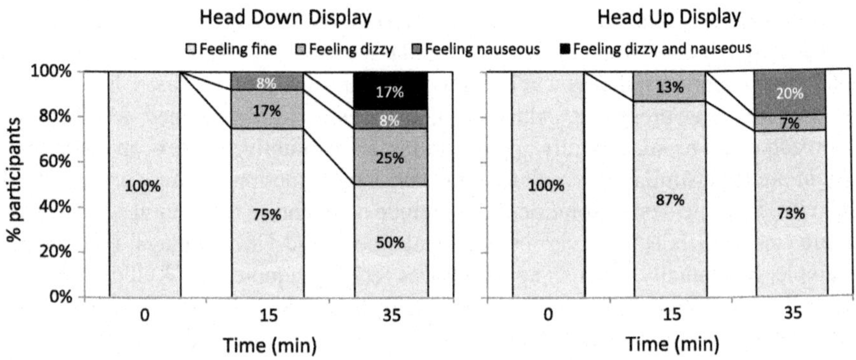

Fig. 1 Percentage of participants reporting motion sickness symptoms during a 35 min drive performing non-driving tasks using a head down display (*left*) and a head up display (*right*) [16]

display showing written text. As expected from a sensory conflict perspective, watching the in-car display led to significantly higher levels of carsickness.

Most recently, Diels et al. [16] reported on a study demonstrating the significance of display position on the occurrence of motion sickness in automated vehicles. Participants were asked to perform various reading and viewing tasks using a tablet located either in the users' lap (head down display) or mounted on top of the dashboard (head up display). The head up display resulted in considerably lower levels of motion sickness (see Fig. 1).

This finding is in line with what would be predicted on the basis of motion sickness theory. In comparison to the head down display, the head up condition enables the user to visually perceive the direction of travel to a larger extent therefore resulting in smaller sensory conflicts and associated motion sickness [17]. In the head down condition, one in four participants reported motion sickness symptoms after 15 min, rising to no less than one in two after 35 min. Positioning the display closer to the line of sight out of the window had a considerable beneficial effect and led to a 50 % reduction in the occurrence of motions sickness. These results clearly illustrate the potential scale of the problem of motion sickness in automated vehicles as well as the importance of designing the driver-vehicle interaction and HMI keeping in mind the causative factors of motion sickness.

Closely related to the above is the observation by Houben et al. [19], who studied the effect of an anti-seasickness display, comparable to that used by Feenstra et al. [11]. As with Diels et al. [16] they differentiated between display position which in this case however showed veridical Earth-fixed self-motion. Their observation was that motion sickness was reduced more when the display was centered on the line of sight than when positioned above it.

At this point, it is appropriate to note that the occurrence of motion sickness can be expected to be closely linked to the vehicle's motion profile. Our organs of balance are in essence biological accelerometers and are subsequently sensitive to accelerations only, i.e., to changes in velocity [18]. As a corollary, sensory conflict as a result of viewing a stationary visual scene is significantly reduced when traveling at constant speed. The organs of balance signal the body to be stationary and any stationary scene as sensed by our eyes will therefore be perceived as congruent. When driving at largely constant velocity, sickness is therefore less likely to occur as a result of reading or using in-vehicle displays. However, the moment dynamic media content is introduced, sensory conflict may of course occur under both constant and varying velocity motion scenarios [2].

3 Design Considerations for Automated Vehicles

From the above, it can be concluded that motion sickness will be of greater concern with automated vehicles compared to conventional vehicles in particular in the light of engagement in non-driving tasks. To minimize the likelihood of motion sickness, there are three fundamental principles that should be observed:

- Avoid vehicle motions around 0.16 Hz
- Allow occupants to anticipate the vehicle's motion trajectory
- Avoid incongruent visual-vestibular self-motion cues.

Although future research will be required to understand how these principles can be suitably applied in the development of future automated vehicles, the three fundamental principles allow us to propose the following initial design considerations [1, 2].

To enable anticipation, window surface areas (also known as Day Light Openings) should be maximized, whereas obstruction by A-pillars, belt or shoulder lines should be minimized. Similarly, seating should be at sufficient height to ensure passengers are able to look out of the window. Fully enclosed cabins and rearward facing seating arrangements can also be expected to exacerbate the problem. Future research may also explore the feasibility of artificial enhancement of the visual scene (e.g. Augmented Reality) possible also displaying the future motion path.

Conflicting motion cues can be minimized by locating displays showing content not related to the outside world near the line of sight out of the window, allowing for peripheral vision to gather information on the direction of travel. Likewise, display size should also be limited to allow for sufficient peripheral visual information and reduce the impact of the visual stimulus. Alternatively, see-through or Augmented Reality displays may avoid the problem of obscuration although issues related to visual comfort may be at stake. Finally, display content (i.e. dynamic vs. static) should be aligned to the vehicle dynamics where possible.

4 Conclusions

Vehicle automation has the potential to provide significant advantages to the driver and society. However, motion sickness may negatively affect the successful acceptance, especially at the critical introductory phase of this technology. Furthermore, it should be acknowledged that motion sickness may have additional consequences [2]. It can compromise task performance and therefore affect the drivers' ability to regain vehicle control. Aftereffects may negatively affect an individual's ability to engage in safety critical activities. Finally, it may prevent the anticipated increase in road capacity if automated vehicle control algorithms need to be tuned to avoid motion sickness. To avoid, or at least limit, the occurrence of motion sickness in automated vehicles and to tackle the problem systematically, it is imperative that we recognize and understand the basic underlying perceptual mechanisms. Future automated vehicles cannot be simply thought of as living rooms, offices, or entertainment venues on wheels.

References

1. Diels C, Will autonomous vehicles make us sick? In: Sharples S, Shorrock S (eds) Contemporary ergonomics and human factors. Taylor & Francis, pp 301–307
2. Diels C, Bos JE Self-driving carsickness. Appl Ergonom (in press)
3. Diederichs F, Bischoff S, Widlroither H, Reilhac P, Hottelart K, Moizard J (2015) Smartphone integration and SAE level 3 car automation – a new cockpit concept and its evaluation in a car simulator. In: Proceedings of the 8th VDI conference Der Fahrer im 21. Jahrhundert [The driver in the 21st Century]. Braunschweig, Germany, Nov 10th–11th, 2015
4. Reason JT, Brand JJ (1975) Motion sickness. Academic Press, San Francisco
5. O'Hanlon JF, McCauley ME (1974) Motion sickness incidence as a function of the frequency and acceleration of vertical sinusoidal motion. AeroSpace Med 45:366–369
6. Guignard JC, McCauley ME (1990) The accelerative stimulus for motion sickness. In: Crampton GH (ed) Motion and space sickness. CRC Press, Boca Raton
7. Bles W, Bos JE, de Graaf B, Groen E, Wertheim AH (1998) Motion sickness: only one provocative conflict? Brain Res Bull 47(5):481–487
8. Le Vine S, Zolfaghari A, Polak J (2015) The tension between autonomous cars' impacts on intersection level-of-service and their occupants' use of travel time for leisurely or economically-productive activities. In: Presented at the 94th annual meeting of the transportation research board
9. Stanney KM, Hash P (1998) Locus of user-initiated control in virtual environments: influences on cybersickness. Presence 7(5):447–459
10. Turner M, Griffin MJ (1999) Motion sickness in public road transport: the relative importance of motion, vision and individual differences. Br J Psychol 90(Pt 4):519–530
11. Feenstra PJ, Bos JE, Van Gent RNHW (2011) A visual display enhancing comfort by counteracting airsickness. Displays 32:194–200
12. Weißgerber T, Dambock D, Kienle M, Bengler K (2012) Erprobung einer kontaktanalogen Anzeige für Fahrerassistenzsysteme beim hochautomatisierten Fahren (Evaluation of a

Contact Analogue Head Up Display for Driver Assistance Systems in Highly Automated Driving. Unpublished Report, Technical University Munich
13. Kyriakidis M, Happee R, de Winter JCF (2015) Public opinion on automated driving: Results of an international questionnaire among 5000 respondents. Transport Res Part F, Traffic Psychol Behav 32:127–140
14. Cowings PS, Toscano WB, DeRoshia C, Tauson RA (1999) The effects of the command and control vehicle (C2 V) operational environment on soldier health and performance (ARL-mr-468). U.S. Army Research Laboratory, Aberdeen Proving Ground, MD
15. Kato K, Kitazaki S (2008) Improvement of ease of viewing images on an in-vehicle display and reduction of carsickness. In: Human factors in driving, seating comfort and automotive telematics, 2008 (SP-2210). SAE Technical Paper Series 2008-01-0565
16. Diels C, Bos JE, Hottelart K, Reilhac P The impact of in-vehicle display position in automated vehicles on the occurrence of motion sickness. Human Factors, submitted
17. Diels C, Bos JE (2015) User interface considerations to prevent self-driving carsickness. In: adjunct proceedings of the 7th international conference on automotive user interfaces and interactive vehicular applications. ACM, pp 14–19
18. Howard IP (1982) Human visual orientation. Wiley, Chichester
19. Houben MMJ, Bos JE (2010) Reduced seasickness by an artificial 3D Earth-fixed visual reference. In: Proceedings international conference on human performance at Sea, HPAS2010, Glasgow, UK, 16–18 June. Turan O, Bos J, Stark J, Colwell JL (Eds) Univ. Strathclyde, Glasgow, UK, 2010, pp 263–270

Potential Solutions to Human Factors Challenges in Road Vehicle Automation

Bobbie D. Seppelt and Trent W. Victor

Abstract Recent research on automated vehicle technologies points to the need to consider drivers' interactions with road vehicle automation, and to apply Human Factors (HF) principles and guidelines to support timely and safe transfer of control to and from automation. This chapter elaborates on a Human Factors breakout session at the 2015 "Automated Vehicles Symposium" that addressed issues on how humans will interact with automated technologies, particularly considering that a wide variety of designs are either under development or already deployed. A number of key human factors design challenges are outlined including that automation is a cost-benefit trade-off where reduced human performance is a cost; that there are different transfer of control concerns for different levels of automation; that the driver may not provide suitable fallback performance of the dynamic driving task; that the better the automation, the less attention drivers will pay to traffic and the system, and the less capable they will be to resume control; and that the driver may be "out-of-the-loop"—may not monitor the driving environment or be aware of the status of automation. Two suggestions to solve the human factors issues are proposed: (1) to work within given constraints, to design the best we can, according to the given definitions of levels 2 and 3 vehicle automation, or (2) to advise against developing level 3 automation and instead advocate two levels of automation: shared driving wherein the driver understands his/her role to be responsible and in control for driving, and delegated driving in which there is no expectation that the driver will be a fallback for performing the dynamic driving task.

B.D. Seppelt (✉)
Touchstone Evaluations, Inc., 18160 Mack Avenue, Grosse Pointe, MI 48230, USA
e-mail: bseppelt@touchstoneevaluations.com

T.W. Victor
Volvo Car Corporation, Research & Development, Volvo Cars Safety Centre, Dept 91400, PV22, 405 31 Göteborg, Sweden
e-mail: trent.victor@volvocars.com; trent.victor@chalmers.se

T.W. Victor
Department of Mechanical Engineering, Chalmers University of Technology, 412 96 Göteborg, Sweden

G. Meyer and S. Beiker (eds.), *Road Vehicle Automation 3*,
Lecture Notes in Mobility, DOI 10.1007/978-3-319-40503-2_11

Keywords Automation · Human factors · Transfer of control · Level 3 automation · Human-automation interaction · Supervised automation · Unsupervised automation

1 Introduction

There is currently an exceptional rate of research and development on automated vehicles, with many prototype systems at all levels of automation being tested on public roads, and some automated functions already on the market (such as adaptive cruise control with lane centering, or automated parking). Human-related research topics such as the role of the human during automation (e.g. driver in or out of the loop), transfer of control, and the need to design HMIs to support system transparency (e.g., communicating system limitations and capabilities) have been brought into focus in the automated vehicle community because of a large body of research pointing out potential problems with automation that are related to the *Human Factor* and driver behavior.

At the 2014 TRB workshop on "Road Vehicle Automation", the Human Factors (HF) breakout group discussed research questions framed around topics of transfer of control from a higher to a lower level of automation or to full manual control, and on the potential for misuse and abuse of automated vehicle technologies. Related to transfer of control, discussions centered on the need to design HMIs to support driver situation awareness and mental model development, and to promote use of systems capable of "failing gracefully". It was proposed that improved feedback on system behavior, either through an HMI or as part of driver training, would address behaviors attributed to unintentional misuse, and constraint of system functionality through forcing functions would address those owing to intentional misuse or abuse.

Recognizing the accelerated pace in research and development of automated vehicles (AVs) and the need for clear direction on issues related to the human, particularly how operators will interact with higher levels of automation, the 2015 Human Factors breakout session set out to identify a set of updated research needs statements. A set of 60 research questions were distilled down to five prioritized ones by over 100 HF professionals in attendance from industry, government, and academia using a modified Delphi method. These five research questions—listed in Table 1—were then reviewed in detail in small groups. Each group produced a draft research needs statement.

Interestingly, four out of the five key human factors research questions in Table 1 have to do with how to provide feedback or information to the driver, and one research question is related to monitoring the driver status. The research questions in Table 1 all speak to the primary underlying concern of how to design AVs to provide a better understanding of the automation and the situation, or conversely to prevent confusion and misuse. Implicitly, these research priorities are

Table 1 Top human factors research questions in the development and deployment of automated vehicles (July 2015)

1. What feedback should automated vehicles provide drivers?
2. How should the car monitor the driver in a test and/or production system?
3. Do we need to communicate the level of confidence of the automation in its decision-making to the driver? How would we do this?
4. How can we indicate to a driver how quickly the system is approaching its boundaries and where the boundaries are?
5. How and when should driving environment information be presented in order to appropriately (re-) orient drivers' attention and awareness back to the roadway?

particularly relevant for Levels 2 and 3 automation [1, 2] as it was assumed in the discussions that development of automation in the near-term would focus on these levels.

1.1 Definitions of Mid-Level Automation—Automation Levels 2–4

Although other taxonomies of levels of automation do exist, the current discussion will refer to the NHTSA [1] and SAE [2] definitions of levels of automation as this was what the HF workshop primarily used.

In *Level 2* automation, the human is still required to participate in the dynamic driving task by monitoring the driving environment and by providing fallback performance of the dynamic driving task [2]. According to SAE [2] the dynamic driving task "includes the operational (steering, braking, accelerating, monitoring the vehicle and roadway) and tactical (responding to events, determining when to change lanes, turn, use signals, etc.) aspects of the driving task, but not the strategic (determining destinations and waypoints) aspect of the driving task". In a similar fashion, NHTSAs defines Level 2 as "integration of braking, throttle, and steering control designed to enable hands free/foot off operation" [1].

In *Level 3*, the human is not required to monitor the driving environment but is expected to respond appropriately to a request to intervene, as a fallback to perform the dynamic driving task according to SAE [2], or similarly in NHTSA's [1] definition the human is expected to be available for occasional control despite giving up full monitoring and control authority.

In contrast, in a *Level 4* system, responsibility for safe operation lies solely with the vehicle and the human is not expected to be available for control even if a human does not respond appropriately to a request to intervene [1, 2]. In Level 4, the system is not designed to rely on the driver as a fallback and responsibility for driving lies with the vehicle, not the driver.

Thus, what differentiates these levels is (a) whether the system is designed so that the driver is expected to *provide fallback performance* of the dynamic driving task (as in Level 2 and 3, but not Level 4) and (b) whether the driver is expected to *monitor the driving environment* (as in Level 2, but not Levels 3 and 4).

1.2 System Limitations

Limitations exist on automation system performance, depending on the level of sophistication of the technology. The core of the matter is that the driver needs to detect these limitations and provide fallback driving performance for situations of limited system performance. For example, early technologies may be sensitive to lane marking visibility (e.g., only working when lane markings are visible), road design (e.g., only working on straight or slightly curved roads), may not detect all collision objects (e.g., pedestrians, animals, or debris), and may be restricted in the amount of force that can be applied in actuation (steering, braking, and acceleration). These limitations may be more or less frequently encountered; for example, loss of lane markings may be more frequently detectable but detectability of animals may be very rarely encountered. The question becomes *what is the best solution to deal with system limitations*?

1.3 Aim

There is an expectation that the driver's role is to provide fallback driving performance and monitor the driving environment (Level 2), or to provide fallback driving performance when requested to intervene, without being required to monitor the driving environment (Level 3). But how do we support drivers to most effectively and safely take back active control of the vehicle (both planned and unplanned transfers)? How do we support drivers in monitoring the driving environment? Is it reasonable to examine the Human Factors research regarding challenges with using the driver as a fallback and the challenges the driver encounters when monitoring the driving environment? This chapter aims at identifying these Human Factors challenges in more detail and aims at providing potential solutions for how to overcome these challenges.

Further, this chapter is also intended to reflect the sentiment of the AVS2015 human factors practitioners, to provide a more detailed summary and discussion on the main human factors lessons of automation, and to provide a perspective from a human factors professional who is actively involved in the design of automated vehicles, leading Volvo Cars safety research on AVs.

The first part of this chapter identifies and provides more detail on key HF lessons of automation from other domains that have deployed automated technology and from existing research in the vehicle domain. This provides a starting point for discussion on the expected benefits and costs of road vehicle automation. This section builds on the human factors research issues identified in AVS2014 and AVS2015. The second part of the chapter discusses potential solutions for the HF challenges.

2 Key Human Factors Challenges

This section identifies and elaborates key human factors challenges in automation, focusing on levels 2 and 3.

2.1 Automation Is a Cost-Benefit Trade-off Where Reduced Human Performance Is a Cost

In other domains it is long-known that automation can both impose a cost and benefit to human performance [3–5]. Automation offers benefits over manual operation with increased efficiency, accuracy, and improved control for routine tasks. Safety and comfort is improved when we automate to alleviate humans from performing difficult tasks and/or tasks that induce boredom, stress, and/or fatigue. In aviation, introduced automation in the cockpit has improved safety, reduced flight times, and increased fuel efficiency. In driving, introduced vehicle control automation promises to improve safety, and improve traffic flow and energy efficiency with eased congestion, greater throughput, and less variability in traffic dynamics.

From a safety perspective, automated technologies, through their advanced sensing, algorithms and crash avoidance systems, have the potential to significantly reduce crashes and save lives. For example, automation has certainly played a role in improving aviation safety where the odds of dying in an airplane crash is 1 in 96,566 compared to 1 in 112 for a motor vehicle crash [6]. This safety improvement is largely because automated technologies are expected to perform better than the human driver, where 94 % of crashes are attributed to driver-related critical reasons such as recognition errors, decision errors, and performance errors [7].

Further, the efficiency benefit alone has the potential to provide significant time and cost savings to commuters. A recent report on urban mobility—Texas A&M Transportation Institute's 2015 Urban Mobility Scorecard study—cited that U.S. drivers lose nearly 7 billion hours each year to traffic congestion—an average of 42 h per commuter—and waste 3 billion gallons of fuel due to these delays with congestion costs estimated at $160 billion [8]. These trends are only expected to increase: by 2020, average delays are projected at 47 h with a total delay climbing to 8.3 billion hours. Such potential to improve routine travel is concomitant with the accelerated push to deploy automation technologies.

Benefits of introduced automation, however, are often derived from those aspects of system operation that do not necessarily consider the interaction with humans, focusing instead on the improved task efficiencies. It is in the interaction with the human that many of the costs of automation arise. Although automation may offload some physical burdens, when systems are imperfectly reliable, operators must monitor the automated system, its performance, and the action that it controls, which leads to cognitive burdens [9, 10]. Potential benefits can be diminished by loss of information due to fundamental changes in system feedback because of automating previously manual tasks. Such feedback changes can lead to operator confusion and reduced awareness of the state and behavior of the automated system [4, 9, 11, 12]. For example, loss of critical haptic, auditory, and visual cues present in manual operation can result in operators having difficulty tracking automation's status and behavior, and a failure to understand when and how to intervene to avoid undesirable actions by the automation or to achieve required performance. Automation can also fundamentally change the feedback operators receive by integrating or processing data in a way that requires interpretation on the part of the operator. This tradeoff of benefits and costs of automation is particularly prominent for imperfectly reliable automated systems—those systems that occasionally require operator intervention due to hardware or software failures, or from when operators use automated systems outside their designed functional limits. Also, partial automation—in which only part of an operator's task is automated—induces the same cost-benefit tradeoff.

The same general pattern of cost-benefit lessons of automation seems to transfer from other domains to vehicle control automation. There are clear performance benefits for routine tasks with use of vehicle control automation, and there are also costs—indications of reduced awareness and capability to recover as vehicle automation increases. Analyses of Adaptive Cruise Control (classified as a level 1 automation) illustrate the cost-benefit relationship. In a recent study on the use of Adaptive Cruise Control (ACC) combined with forward collision warnings (FCW) in a field operational trial (FOT) under normal driving conditions, a positive safety effect was observed showing a reduced number of harsh braking events, less critical time headways (THWs; those <0.5 s), and less incident events (as defined by video and kinematic triggers) as compared to periods of manual control [13]. This net positive effect, which was attributed to increased safety margins (time headway), was present despite there being a general increase in secondary task involvement [13] and an increase in eyes off path [14]. Thus, there was a net positive safety benefit. This indicates that, when assessing the overall impact of automation, a holistic approach should be taken. Safety effects should be considered at a joint driver-vehicle control system level, whereby positive effects from automation (such as increased time headway and increased lateral protection) can offset potential negative effects such as increased secondary task engagement. This implies that the smallest unit of analysis is the driver and vehicle in a given context.

2.2 *Different Transfer of Control Concerns for Different Levels of Automation*

Transfer of control to the driver refers to those situations when the driver must resume control of the dynamic driving task. According to SAE [2], the dynamic driving task "includes the operational (steering, braking, accelerating, monitoring the vehicle and roadway) and tactical (responding to events, determining when to change lanes, turn, use signals, etc.) aspects of the driving task, but not the strategic (determining destinations and waypoints) aspect of the driving task". The transfer of control is either system-initiated or driver-initiated. A system-initiated transfer occurs when the driver receives a request to take-back control due to limitations of the automated system to manage a particular driving situation or environmental condition. A system-initiated transfer can be planned (e.g., exiting an automation-suitable road section) or unplanned (e.g., a malfunction). The driver may self-initiate a transfer in anticipation of the system's approach to its functional or design limits (e.g., approaching sharp curved sections of road) due to some level of discomfort or discontent with the dynamics of the system's response to the driving environment (e.g., inappropriate positioning in relation to merging vehicles), to fulfill tactical goals (e.g., lane changes to avoid slower moving traffic), or to fulfill navigation goals (e.g., exiting a motorway with infrastructure support to reach a destination in an unsupported urban or rural environment [15]).

In the discussion of transfer of control back to the human operator, there is an important distinction of issues between Level 2 and Level 3 systems. For NHTSA Level 2 systems (combined function automation), while activated, the automated driving system performs lateral and longitudinal control functions (ACC + lane keeping assistance) but the driver is required to monitor the driving environment and respond to unexpected objects or events. The driver is expected to be "available for control at all times and on short notice" and "to be ready to control the vehicle safely" as "the system can relinquish control with no advance warning" [7] Automated systems at level 2 are *designed and intended to be support systems* to help the driver manage vehicle control. Technologies such as ACC in combination with lane centering, while alleviating the driver from physically operating the vehicle, still require intermittent-to-frequent input from the driver; for example, drivers must provide added steering torque in curves when using lane centering systems and are prompted to periodically return their hands to the steering wheel during straight sections of roadway because lane markings can be lost due to unexpected roadway conditions or poor sensing quality. The primary HF concern at level 2 is the need to calibrate driver expectation to system capability.

Because Level 2 systems require the driver to resume control on a moment's notice, a concern for this level of automation is if drivers will be able to maintain sufficient situation awareness without continuous active engagement in (i.e., moment-to-moment) vehicle control to safely and effectively perform an infrequent hazard detection task. Key HF design issues are the provision of sufficient feedback to ensure appropriate reliance on system control, to minimize secondary task involvement, and to prevent mode confusion where the driver assumes the automation is more capable than it actually is. Concerns over a driver's ability to resume control are more pronounced for Level 3 systems. For NHTSA Level 3 systems (limited self-driving), while activated, the automated driving system performs the complete dynamic driving task, including lateral and longitudinal control functions, as well as object and event detection and response. While the vehicle is "designed to ensure safe operation during automated driving mode", the driver is not expected to monitor the driving environment but is expected "to be available for occasional control" and to respond "appropriately", provided "sufficiently comfortable transition time", in the event of a hand-off of the dynamic driving task from the automated driving system to his/her manual control [7]. Given this requirement to re-engage in the vehicle control loop (at a sufficiently comfortable transition time), a key concern for these systems is if this is compatible with human performance and whether instead the human driver should in fact be required to monitor the driving environment to some extent, such as in a requirement to monitor the vehicle's response to the driving environment. Level 3 also raises HF concerns regarding limitations that drivers have in performing vigilance tasks (i.e., monitoring tasks with infrequent control activity), and timely resumption of control due to expected increases in drivers' secondary task engagement during periods of automated control. A key HF design issue at this level is how to design transfer of control requests, both in terms of their timing relative to required manual control periods and in their presentation modality/frequency.

2.3 The Driver May not Provide Suitable Fallback Performance of the Dynamic Driving Task

Planned transfers of control such as those in Level 3 automation should occur on a timescale that provides drivers a "sufficiently comfortable transition time" [1]. Recent research from Gold and colleagues [16] suggests that a minimum of 5–7 s in advance of the required period of manual driving is needed to engage the attention of the driver presuming the driver is out of the loop (i.e., not actively monitoring the driving environment). Note that the 5–7 s takeover request time window is not feasible in a Level 2 system. This lead time is a minimum requirement to engage drivers as at 5-s prior to the occurrence of the system boundary, though sufficient to avoid any collisions, drivers enacted suboptimal control maneuvers. As part of a high-fidelity driving simulator study, drivers were alerted to the need to take-back control from a highly-automated vehicle (capable of performing lateral and longitudinal control as well as lane changes and overtaking maneuvers up to a maximum speed) either 5 or 7 s in advance of an accident in their lane of travel, which required either evasive steering or a brake response to avoid [11]. They were instructed to either put their hands back on the steering wheel or to press the brake pedal to take back vehicle control. As compared to the 7-s response lead, with the shorter 5-s lead time, gazes to mirrors and shoulder checks decreased, accelerations

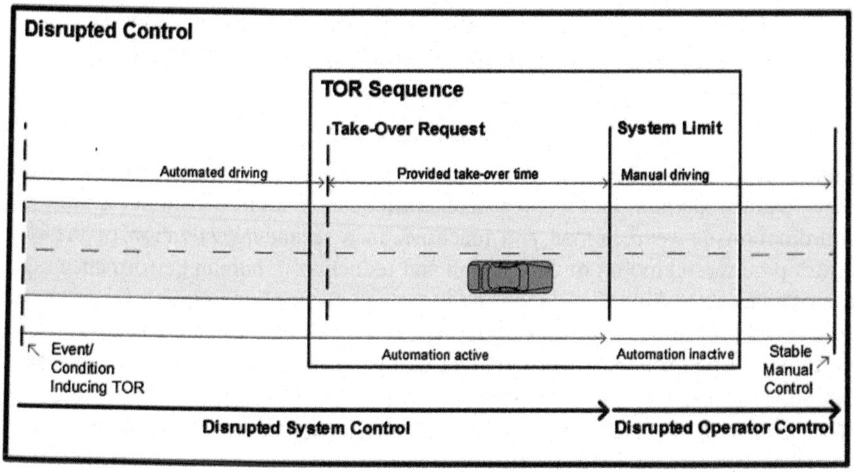

Fig. 1 Longer envelope of disrupted control surrounding take-over requests in a Level 3 automation system (adapted from 17)

potentials increased, and the brake was overused at the expense of lateral maneuvering. Compared to a baseline driving group, in which they drove manually and received no advance alert, drivers in the automated conditions generated acceleration potentials close to three times higher and performed more sudden and intense braking maneuvers. These results call into question if even seven seconds provides sufficient time for drivers to enact a "safe" take-over. Added to a take-over-request time window for planned transfers of control are the surrounding disruptive effects on vehicle and driver control. As shown in Fig. 1 surrounding the chronological sequence of a take-over process (or TOR sequence; [17]) with a transition from a highly-automated to manual driving is an upstream disruption of system control and a downstream disruption of driver control.

Prior to the system's request for a driver to resume control in Level 3 automation, events or conditions necessarily push automation to move towards its operating boundaries. Following from take-over, drivers show a period of degraded control, in which they require 10 s (if predictable) to 35 s (if unpredictable) to stabilize their lateral control of the vehicle [15]. Such disruption advocates for HF methods or design considerations to help ensure drivers develop appropriate expectations of the vehicle's capabilities and response behavior, both for planned and unplanned transfers of control. When this longer envelope of disrupted control is considered, effects of transfers of control may extend to minutes. Humans are challenged to remain vigilant to anticipate transfers of control, which would presumably enable them to reduce their disruptive effects. And when alerted to the need to take-over control, there are notable impracticalities for expecting drivers to initiate a timely and safe intervention response. Here too it is worth note that varying levels of driver response capabilities and task engagement at the time of the request will likely further hamper a seamless take-over response.

2.4 *The Better the Automation, the Less Attention Drivers Will Pay to Traffic and the System, and the Less Capable They Will Be to Resume Control*

Imperfectly reliable automation and partially automated tasks are similar in that they require operator involvement and consequently require communication and coordination between human and machine. In a recent investigation of the relationship between amount of automation and reduction in human performance (such as complacency, skill degradation, and loss of situation awareness), Onnasch and colleagues [18] found that while an increased amount of automation support results in improved performance for routine tasks, operators have a reduced awareness of the situation or operating environment, and show difficulty troubleshooting and recovering if something goes wrong with the automation or if something unexpected happens. This finding—of reduced awareness and capability to recover as automation increases—is based on a meta-analysis of 18 studies from process control, supervisory control, and aviation, and is robust across domains. Such automation-induced performance consequences are largely attributed to operators' tendency to *reduce their monitoring of highly reliable automation because of its ability to function properly for an extended period of time* [19, 20]. In mid- or medium levels of automation, in which the operator may be required to resume manual control, the prevailing take-away lesson is to keep operators 'in the loop', either through their involvement to some extent in decision and action selection tasks as well as action implementation [21] or through intuitive, "ecological" displays on the state of the automated processes [22–24].

2.5 *The Driver May Be "Out-of-the-Loop"—May not Monitor the Driving Environment or Be Aware of the Status of Automation*

Introducing vehicle automation does not simply relieve drivers of tasks and replace drivers with a more reliable vehicle control system. Instead, it introduces new tasks that the driver must do such as configuring, engaging, and monitoring the automation [3, 9]. Human-automation issues are likely to arise if the human does not understand the automation (in terms of capabilities, boundaries of operation, current functionality, goals, and level of automation [25]). Spanning level 2 and 3 of automated driving is the dilemma of how to keep drivers sufficiently 'in the loop' when interacting with automated system(s) so they can intervene if necessary yet still provide the full benefits and conveniences promised with these mid-levels of automation. The inherent attentional constraints when performing vigilance tasks and distraction tendencies when moment-to-moment control is not required are actively working against a driver's ability to seamlessly resume control. Implicit in a driver's task of monitoring the driving environment is his/her ability to scan for intermittent, unpredictable, and infrequent or rare events that may require a control response—essentially, to perform a vigilance task. Arguably, many of the situations that cause crashes are unexpected and rare events. Humans are not known to be particularly effective in this role. In other domains, example vigilance tasks include monitoring for infrequent contacts (radar monitor), examining x-rayed carry-on luggage (airport security inspector), and observing a stream of products to detect and remove defective or flawed items (quality control inspector). Analyses of operators' performance for these tasks generally conclude that (1) operators frequently show lower vigilance levels (defined as steady-state level of vigilance performance) than desired or required to adequately detect events or signals, and (2) the vigilance level commonly declines steeply during the first half hour or so of a watch [26]—an effect of enough prominence to have a term associated with it—"vigilance decrement". More than five decades of research studying humans in vigilance tasks [27–30] point to the need to supplement operators' ability to remain attentive either through training paradigms [31], periodic manual control [32] or display techniques [33, 34]. In terms of distraction tendencies, drivers seem quick to take advantage of the reduced demand automated systems afford in their assumption of moment-to-moment vehicle control in that they are inclined to direct their attention away from the forward roadway to other locations. Recent driving simulation studies that have examined how highly automated technologies (ACC, LKA, and ACC + LKA) affect driver attention to road center have shown that *drivers' attention decreases as the level of automation increases*, and that the type of automation support provided to drivers (lateral versus longitudinal) results in different levels of driver engagement and performance [35–37]; automating only lateral control produces scan behavior similar to when both lateral and longitudinal control are automated, with significantly less attention paid to the forward road as compared to conditions with use of automated longitudinal control or manual control. Reduced scanning of the forward roadway (likely a consequence of an increased uptake of secondary task activity) results in a reduced awareness of the surrounding traffic and roadway conditions. This shift in attention can be detrimental to driving safety, particularly in instances when there is an obligation to resume control of driving, e.g., to change lanes due to an incident on the roadway [38]. As a direct performance consequence, drivers can be slower to respond to a warning of critical events [38, 39] when both lateral and longitudinal tasks are under automated control.

Arguably, concerns over insufficient driver vigilance and distraction tendencies are constrained to instances when the automated system or other vehicle safety systems on-board fail to alert the driver of the need to resume control—i.e., *unplanned transfers*. However, even with planned transfers, in which drivers receive forewarning of the need to resume control, as discussed earlier, there are some concerning practical realities of the required length of a take-over time window to support timely, effective resumption of driver control, as well as overlooked issues of degraded control that extend this time window and its degraded effects on the combined driver-vehicle performance. These situations—when silent failures cause the system to act dangerously—may be the most concerning issue with an automation solution that does not require the driver to monitor the driving environment but be available for prompted, planned transfers of control (Level 3 automation). Clearly, silent failures are incompatible with Level 3 system solutions. If a failure does occur, operators may require a significant period of time to reorient themselves to the current system state and to develop adequate understanding of the state to act appropriately, thus potentially reducing the effectiveness of actions taken and prohibiting operators from carrying out required actions. For example, drivers removed from active steering and speed control with active steering (AS) and adaptive cruise control (ACC) systems, respectively, have been shown to fail to effectively intervene in response to failures, largely a result of missing haptic feedback in control of the steering wheel and of raw data on the automation's low-level processing in control of speed [12]. In essence, drivers are at risk for out-of-the-loop unfamiliarity when they are removed from the vehicle control loop and unsupported in their monitoring role. Driver *out-of-the-loop unfamiliarity* (OOTLUF) is defined as a decreased ability to detect system errors and to intervene and perform the task(s) manually in response to failures compared to those who manually perform the same task or set of tasks [5, 10, 25]. Out-of-the-loop unfamiliarity stems from disrupted feedback that diminishes the ability to form correct expectations, detect anomalies, and control the system manually. It is the qualitative shift from active behavior to passive observation and its associated loss or change in the type of feedback received on the state of the system that can lead drivers to develop inaccurate mental models, over- or under-trust the automation respective to its capabilities, display complacent behavior, and adapt their behavior in safety-degrading ways [9, 11, 40, 41]. Indicators of OOTLUF, more broadly, include:

- Delayed response time (RT) to system failures (either lack of response—missing control—or mis-calibrated or inaccurate response to system failure).
- Complacent reliance—refers to a driver's self-satisfaction that may result in a lack of vigilance based on an unjustified assumption of satisfactory system state. Overreliance on a driving automation system is sometimes termed complacency when it results from trusting a system more than is warranted [42].
- Inaccurate mental models (as measured subjectively by testing knowledge of actions and limits of the system, i.e., its boundary conditions). *Note*: Operators with substantial previous experience and well-developed mental models detect disturbances more rapidly than operators without this experience [43].
- Reduced manual control performance.
- Inaccurate or incomplete expectations of system response and behavior, i.e., inability to anticipate situations that lie beyond the capabilities of the automation.
- Passive monitoring (glance behavior—e.g., failure to sample safety-critical areas such as crosswalks at intersections, or to make glances to rear-view mirror and/or side mirrors or over-the-shoulder prior to lane changes [16]).

- Increased uptake of secondary tasks.
- Unnoticed mode transitions [44].
- Low situation awareness scores or loss of awareness of the state and processes of the system [25].
- Lower reported self-confidence scores to make decisions (or control vehicle manually) after system failure.
- High trust scores indicative of over-trust based on system capability for response—also evident in delayed or lack of response to warnings, e.g., for ACC, delayed and less effective braking responses in situations in which the ACC is not able to respond to a braking lead vehicle [12].

The prevailing HF concern is how to reduce or prevent these emotional and behavioral consequences for drivers that can result when the moment-to-moment control tasks are automated yet they are still required to respond appropriately to requests to intervene for NHTSA and SAE Levels 2 and 3.

3 Potential Solutions for the Human Factors Challenges in Mid-Level Automation

Taking a step back and looking at the research results presented in the previous section, it would appear that Human Factors research is advising against higher levels of automation, in particular level 3 automation that assumes the driver can be a fallback solution despite not monitoring. Results indicate that the better automation becomes, the more the driver becomes out of the loop, the less capable s/he is to recover. On the other hand, imperfect automation is to be compared with the alternative of manual driving by a driver who is far from perfect with 94 % of crashes being attributed to driver-related critical reasons such as recognition errors, decision errors and performance errors [7]. It seems to be a classic dilemma, if we don't automate we are stuck with the human contribution to crashes, but if we do choose to automate, human performance will get worse as automation gets better (!). This dilemma has been recognized by some Human Factors researchers, such as Norman [45], but clearly there has to be a plan to resolve this dilemma.

A number of alternatives are presented below as possible approaches to mitigate the human factors challenges with level 2 and level 3 automation as outlined in Sect. 2 above.

3.1 Alternative 1: Work Within Given Constraints—Design the Best We Can, Given the Definitions of Level 2 and 3

One approach to solve or mitigate the HF challenges is to make the best of it and work within the current constraints and definitions of levels 2 and 3 of automation (see the levels definitions in Sect. 1 above). For example, to accept that automation is designed with the driver as a fallback for automation, despite not being required to monitor the driving environment.

It does appear that the top five HF research questions selected in the AVS2015 match up well as research proposals to make the best of it, to improve or mitigate most of the human factors challenges identified in Sect. 2 within the given system constraints, primarily by providing better feedback and attention-orienting assistance. Three of the five questions have to do with improving feedback about the vehicle automation status (i.e. improved feedback in general, providing decision-making confidence, and informing about approaching system operating boundaries), one question addresses how to provide driving environment information to reorient the driver's attention to the roadway, and one question was related to monitoring the driver's status. Thus, the top research questions in Table 1 all address the primary concern of how to provide a better and timely understanding of the automation itself and the situation. Much research is currently ongoing regarding development of best practices and design guidelines for the design of road vehicle automation [46, 47].

Shared control approaches have been advocated [48, 49], in which it is suggested that the human should always remain in control, but should be able to experience or initiate smooth shifts between levels of authority [48]. This approach seems excellent when the driver is expected to monitor the traffic environment and partake in the control of the dynamic driving task with the support of the automation (as in a Level 2 system); however, it does seem completely incompatible with level 3 automation which allows the driver to withdraw from monitoring the driving environment. That is, it changes the definition of a level 3 system to require the driver to monitor the driving environment, effectively changing a level 3 system into a level 2 system.

It must be stated that imperfect, partial automation may still be safer than today's imperfect drivers, despite the many unsolved HF challenges [45, 50]. It could be the case that level 3 automation is necessary to pass before achieving higher levels of automation, but then again maybe not.

3.2 *Alternative 2: Advice Against Level 3—Advocate for Two Levels of Automation (Shared and Delegated Driving)*

Considering the fairly conclusive results on human monitoring and response deficiencies outlined above, an alternative solution is to *advise against the level 3 solution* which is designed with the human as a fallback to perform the dynamic driving task – a human that is not required to monitor the driving environment but is expected to respond appropriately to a request to intervene.

Instead, a conclusion could be made from the review that we need to polarize the "levels of automation" into two types of automation—shared driving (or supervised Level 1 or 2 automation) and delegated driving (or unsupervised Level 4 or 5 automation). In the current context this means that this suggestion is to intentionally stay at level 2 automation until such a point where level 4 automation can be achieved. In shared driving (level 2), the role of the driver should be clear. The driver is responsible for driving and is receiving support for this role from the automation. In level 2, the role of automation is to support the driver's actions, much like a manager-employee relationship where the driver is the manager and the automation is the employee. Thus, we should emphasize the driver's role as being in command but having support to perform better.

Whereas in delegated driving (unsupervised automation or levels 4 and 5) the automated system is *not* designed to be reliant on the human as a fallback, because it is evident from research that the driver does not function well as a fallback solution. When automation is designed so that it cannot use the driver as a fallback, there are other precautions that will be taken such as system redundancy and fulfillment of higher levels of ISO 26262 functional safety.

4 Conclusions

This chapter started by outlining a number of key human factors design challenges:

- that automation is a cost-benefit trade-off where reduced human performance is a cost; that there are different transfer of control concerns for different levels of automation;
- that the driver may not provide suitable fallback performance of the dynamic driving task;
- that the better the automation, the less attention drivers will pay to traffic and the system, and the less capable they will be to resume control; and
- that the driver may be "out-of-the-loop"—may not monitor the driving environment or be aware of the status of automation.

When we take drivers out of the vehicle control loop by automating primary vehicle control tasks we place them in the role of supervising the automation and its performance—fundamentally a vigilance task—a task known to undercut timely and effective human response. Review of research on a driver's ability to act in this role and to serve as a fallback performer point to inherent concerns with Level 3 systems in particular—those systems for which the driver is both taken out of the control loop yet still held responsible for staying aware of the driving environment. When designed well, despite removing drivers from the primary vehicle control tasks, Level 2 systems can offer a performance benefit and improve overall driving safety because they still engage the driver to be responsible for driving and to monitor the driving environment (e.g. for system limitations such as missing lane markings). In order to be successful, however, the key will be not just the development of the automated technology (i.e., reliable sensing capability) but duly to support the driver to be in the loop and to inform the human of its state, limitations, and capabilities through its HMI.

Addressing how to design these systems to achieve these aims is the key to navigate through these mid-levels of automation to when Level 4 systems are realizable. But in the interim, as we navigate through partial automation, given what we know about the human factor in response to automated control, it is imperative that we avoid the pitfalls in expecting humans to compensate for poorly designed or insufficiently-capable automation.

Two suggestions to solve the human factors issues are proposed: (1) to work within given constraints, to design the best we can, according to the given definitions of level 2 and 3, or (2) to advise against developing level 3 automation and instead advocate two levels of automation: shared driving wherein the driver understands his/her role to be responsible and in control for driving, and delegated driving in which there is no expectation that the driver will be a fallback for performing the dynamic driving task.

References

1. NHTSA (2013) US department of transportation policy on automated vehicle development, p 4
2. SAE (2014) Taxonomy and definitions for terms related to on-road motor vehicle automated driving systems. SAE Standard J3016, USA
3. Bainbridge L (1983) Ironies of automation. Automatica 19(6):775–779
4. Kessel CJ, Wickens CD (1982) The transfer of failure-detection skills between monitoring and controlling dynamic systems. Hum Factors J Hum Factors Ergon Soc 24(1):49–60
5. Wiener EL, Curry RE (1980) Flight-deck automation: promises and pitfalls. Ergonomics 23 (10):995–1011
6. NSC (2015) Odds of dying 2015. National Safety Council. Available: http://www.nsc.org/act/events/Pages/Odds-of-Dying-2015.aspx
7. NHTSA, Critical reasons for crashes investigated in the national motor vehicle crash causation survey, A brief statistical summary, DOT HS 812 115
8. Schrank D, Eisele B, Lomax T, Bak J (2015) Urban mobility scorecard, the texas A&M transportation institute & INRIX, Inc., Aug 2015
9. Sarter NB, Woods DD, Billings CE (1997) Automation surprises. In: G. Salvendy (ed.) Handbook of human factors & ergonomics, second edn. Wiley
10. Wickens CD (1992) Engineering psychology and human performance, 2nd edn. Harper Collins, New York
11. Norman DA (1990) The problem of automation: inappropriate feedback and interaction, not "over-automation". In: DE Broadbent, A Baddeley, JT Reason (eds.), Human factors in hazardous situations. Oxford University Press, Oxford pp 585–593
12. Stanton NA, Young MS (1998) Vehicle automation and driving performance. Ergonomics 41 (7):1014–1028
13. Malta L, Ljung Aust M, Gustafsson D, Dobberstein J, Benmimoun M, Pütz A, Saint Pierre G, Metz B, Freek F (1998) euroFOT deliverable 6.4, Final results—Impacts on traffic safety, Brussels

14. Tivesten E, Morando A, Victor T. (2015) The timecourse of driver visual attention in naturalistic driving with adaptive cruise control and forward collision warning, Paper presented at Driver Distraction and Inattention
15. Merat N, Jamson H, Lai F, Daly M, Carsten O (2014) Transition to manual: driver behavior when resuming control from a highly automated vehicle. Transp Res Part F 27:274–282
16. Gold C. Dambock D. Lorenz L, Bengler K (2013) Take over!"—How long does it take to get the driver back into the loop? In: Proceedings of the Human Factors and Ergonomics Society, 57th annual meeting, 2013, 1938–1942,
17. Damböck D, Farid M, Tönert L, Bengler K (2012) Übernahmezeiten beim hochautomatisierten Autofahren. 5. Tagung Fahrerassistenz 2012. München. Germany
18. Onnasch L, Wickens CD, Li H, Manzey D (2013) Human performance consequences of stages and levels of automation, an integrated meta-analysis. Hum Factors J Hum Factors Ergon Soc
19. Parasuraman R, Molloy R, Singh IL (1993) Performance consequences of automation-induced 'complacency'. Int J Aviation Psychol 3(1):1–23
20. Yeh M, Merlo JL, Wickens CD, Brandenburg DL (2003) Head up versus head down—the costs of imprecision, unreliability, and visual clutter on cue effectiveness for display signaling. Hum Factors J Hum Factors Ergon Soc 45(3):390–407
21. Parasuraman R, Sheridan TB, Wickens CD (2000) A model for types and levels of human interaction with automation. Syst, Man and Cybern, Part A: Syst umans, IEEE Trans 30(3): 286–297
22. Bennett KB, Flach JM (2011) Display and interface design: subtle science, exact art. FL, CRC Press, Boca Raton
23. Burns CM, Skraaning G, Jamieson G, Lau N, Kwok J, Welch R, Andresen G (2008) Evaluation of ecological interface design for nuclear process control: Situation awareness effects. Hum Factors 50:663–698
24. Seppelt BD, Lee JD (2007) Making adaptive cruise control (ACC) limits visible. Int J Hum Comput Stud 65:192–205
25. Endsley MR, Kiris EO (1995) The out-of-the-loop performance problem and level of control in automation. Hum Factors 37(2):381–394
26. Wickens CD, Hollands JG (2000) Engineering psychology and human performance. NJ, Prentice Hall Inc, Upper Saddle River
27. Mackworth NH (1948) The breakdown of vigilance during prolonged visual search. Quarterly Journal of Experimental Psychology 1:5–61
28. Davies DR, Parasuraman R (1982) The psychology of vigilance. Academic Press, London
29. See JE, Howe SR, Warm JS, Dember WN (1995) Meta-analysis of the sensitivity decrement in vigilance. Psychology Bulletin 117:230–249
30. Warm JS (ed) (1995) Sustained attention in human performance. Wiley, Chichester
31. Fisk AD, Schneider W (1981) Controlled and automatic processing during tasks requiring sustained attention. Hum Factors 23:737–750
32. Parasuraman R, Mouloua M, Molloy R (1996) Effects of adaptive task allocation on monitoring of automated systems. Hum Factors 38(4):665–679
33. Luzzo J, Drury CG (1980) An evaluation of blink inspection. Hum Factors 22:201–210
34. Lewandowski LJ, Kobus DA (1989) Bimodal information processing in sonor performance. Human Performance 2:73–84
35. Carsten O, Lai FCH, Barnard Y, Jamson AH, Merat N (2012) Control task substitution in semi-automated driving, does it matter what aspects are automated? Hum Factors 54:747–761
36. Jamson H, Merat N, Carsten O, Lai F (2011) Fully-automated driving: the road to future vehicles. In: Proceedings of sixth international driving symposium on human factors in driver assessment, Training and Vehicle Design
37. Jamson A.H, Merat N. Carsten O, Lai FCH (2013) Behavioral changes in drivers experiencing highly-automated vehicle control in varying traffic conditions. Transp Res C Emerg Technol 116–125
38. Merat N, Jamson H, Lai F, Carsten O (2013) Highly automated driving, secondary task performance and driver state. Hum Factors 54:762–771

39. Merat N, Jamson AH (2009) How do drivers behave in a highly automated car. In: Proceedings of the 5th international driving symposium on human factors in driver assessment, Training and Vehicle Design, pp 514–521
40. Lee JD, Moray N (1994) Trust, self-confidence, and operators' adaptation to automation. Int J Hum Comput Stud 40(1):153–184
41. Parasuraman R, Mouloua M, Molloy R (1994) Monitoring automation failures in human-machine systems. Current research and trends, Human performance in automated systems, pp 45–49
42. Ward NJ (2000) Automation of task processes: an example of intelligent transportation systems. Hum Factors Ergonomics 10(4): 395–408
43. Wickens CD, Kessel C (1981) Failure detection in dynamic systems. In: Human detection and diagnosis of system failure, Springer US, pp 155–169
44. Sarter NB, Woods DD (1995) How in the world did we ever get into that mode? Mode error and awareness in supervisory control. Hum Factors J Hum Factors Ergonomics Soc 37(1):5–19
45. Norman DA (2015a) The human side of automation. In: Meyer G, Beiker S (eds) Road vehicle automation 2, Springer International Publishing
46. Kelsch et al. (In Prep). AdaptIVe deliverable D3.3, final functional HF recommendations, EU project FP7-ICT-2013.6.5. www.adaptive-ip.eu
47. Campbell JL, et al. Human factors design assistance for level 2 and level 3 automated driving concepts, Draft version (In Press)
48. Abbink DA, Mulder M, Boer ER (2012) Haptic shared control: smoothly shifting control authority? Cognition, Technology & Work, 14:19–28. http://doi.org/10.1007/s10111-011-0192-5
49. Flemisch FO, Bengler K, Bubb H, Winner H, Bruder R (2015) Towards cooperative guidance and control of highly automated vehicles—H-Mode and Conduct-by-Wire. Ergonomics 57 (3):343–360
50. Norman DA (2015b) Automatic cars or distracted drivers—We need automation sooner, Not later. http://www.jnd.org/dn.mss/automatic_cars_or_di.html

Part III
Ethics, Energy and Technology Perspectives

Connected Autonomous Vehicles: Travel Behavior and Energy Use

Jonathan Rubin

Abstract Autonomous vehicles offer great promise for unprecedented improvements in mobility and safety. However, self-driving vehicles may also significantly alter behavior because they can make driving easier and safer. This may lead connected autonomous vehicles to have large unintended consequences in terms of additional energy use and greenhouse gas emissions, as well as causing decreases in the density of urban areas and may impact congestion. This paper uses consensus estimates from the literature on the cost of driving and the value of travel time to evaluate automation's ability to reduce the costs of travel time. Policy solutions to address the induced driving include charging for miles driven taking into account when and where vehicles are used.

Keywords Autonomous vehicles · Energy use · Travel costs · Economics

1 Introduction

Automated vehicles offer great promise for unprecedented improvements in mobility and safety. In anticipation of this technology, a growing number of US states have enacted legislation that permits operation of self-driving vehicles under restricted conditions designed to determine their use and compatibility on public highways [1] These experimental vehicles are at the higher end of a wide range of automation that begins with some safety features already in vehicles, such as electronic stability control and antilock braking systems. The US National Highway Traffic Safety Administration [2] defines vehicle automation as having five levels from no automation (Level 0) where the driver is in complete and sole control of the primary vehicle controls, to full self driving automation (Level 4). The Society of Automotive Engineers has a similar, classification system but splits Level 4 into two categories,

---

J. Rubin (✉)
Margaret Chase Smith Policy Center and School of Economics,
University of Maine, 5782 Winslow Hall, Orono, ME 04469, USA
e-mail: rubinj@maine.edu

G. Meyer and S. Beiker (eds.), *Road Vehicle Automation 3*,
Lecture Notes in Mobility, DOI 10.1007/978-3-319-40503-2_12

one mode-specific (e.g., highway) and one that can operate in all modes and weather conditions [3]. Currently, automated features including adaptive cruise control and lane keeping are being introduced on premium model lines of various manufacturers. These are generally considered Level 2 automation since they involve "automation of at least two primary control functions designed to work in unison" but still require drivers to monitor and be responsible for the vehicle at all times.

The higher levels of automation, Level 3 and above have the vehicle performing all aspects of driving such that drivers are able, under a variety of circumstances depending on level of automation, to perform other tasks. This aspect of automation is likely to significantly alter driver behavior because it allows for a more enjoyable driving experience where the driver can engage in other activities. Additionally, high levels of automation may enable higher speeds that cause road travel to be more competitive with aviation (for passengers) and rail (for freight).

The higher levels of automation require automated vehicles to be connected to real-time route mapping software and allow for optimizing road situational awareness such as anticipating congestion and construction obstacles. While connectivity enhances or is required for AVs, connectivity is also increasingly becoming integrated with new vehicles that will not have higher levels of automation. Thus, from a behavioral perspective, connected highly automated vehicles (CAVs) differ from connected vehicles (automation Levels 1 or 2) in their ability to allow drivers to perform other tasks, but do not necessarily have advantages in terms of optimal route finding.

Due to the behavioral responses of drivers, CAVs may have large unintended consequences in terms of additional energy use and GHG emissions in the transportation sector and may also have the potential to further decrease the density of urban areas. This paper uses consensus estimates from the literature on the cost of driving and the value of travel time to estimate the potential increases in driving due to automation. Policy solutions to address the induced driving are discussed. The barriers to meaningfully reduce the impact of driving are also acknowledged.

2 Literature Review

There is a small literature that tries to look at the likely impacts of CAVs on driving and energy use [4–8]. As a whole they tend to be optimistic about the benefits of the CAV technology. Mackenzie et al. bound the long-term (2050) energy implications of road vehicle automation using a Kaya Identity (ASIF) based on the framework of Shipper [9] where emissions and energy are given by:

$$Emissions = \boldsymbol{Activity\,Level} \cdot \boldsymbol{Modal\,Share} \cdot \boldsymbol{Energy\,Intensity} \cdot \boldsymbol{Fuel\,Carbon\,Content}$$

They identify a number of ways in which Level 4 automation may have a significant impact on many important aspects of personal vehicle use. These potential impacts include (in part): reductions in congestions, improved crash avoidance, higher highway speeds, vehicle platooning, traffic smoothing (efficient driving), improved routing, reduction in the time and insurance costs of ownership and the enabling of new user groups such as the elderly and the disabled. As noted, CAVs can also potentially increase energy and emissions by inducing additional and faster driving, and by drawing in underserved populations (disabled and elderly drivers).[1] Because of the inherent uncertainties, Mackenzie et al.'s estimated range of impacts on transportation travel demand, energy use and GHG emissions, is very large, −40 to +100 %. This range becomes smaller if only considering changes due to automation, not vehicles, vehicle costs or improved routing from increases in connectivity.

Fagnant and Kockelman [8] note that shared autonomous vehicles (SAV) may increase the use of car sharing programs such as Car2Go and Zipcar by reducing the barriers, including users' need to travel to access available vehicles. Their simulation results show that, in an urban setting, each SAV can replace around eleven conventional vehicles, but adds up to 10 % more travel distance than comparable non-SAV trips, resulting in overall beneficial emissions impacts. It is unclear, however, how much of this savings is due to lower per vehicle emissions of the SAVs, which are all sedans, as compared to the US average light-duty vehicle fleet that includes light-duty trucks and SUVs.

What is important to understand from a perspective of the realized impacts of CAVs are the factors that influence how they will be used and by whom. While potential vehicle efficiency improvements and other system optimizations matter, the largest impacts are likely to come from how consumers react to automation in terms of use: how often, when and how much they use their vehicles. This is largely determined by the full cost and value to drivers who decide to take a trip by car, use public transportation or walk. CAVs will only enter the market place and be privately purchased because they enhance driving. Clearly, enhanced safety and access to mobility by users limited by age or ability is an unmitigated benefit. Beyond this, increasing amounts of automation will only be purchased because they will provide drivers with an improved quality of travel time, allow for drivers to participate in non-driving activities and may enhance drivers' perceived sense of social status.

3 Costs of Driving

The total cost of operating a vehicle varies depending on the type of vehicle, when and where the vehicle is used and for what purpose (work or leisure). The marginal cost for an additional mile includes fuel, tire wear, tolls and the time and pleasure or displeasure of the driver while driving. The average cost for some length of period

---

[1]A number of these papers also identified other potential impacts such as changes in vehicle performance and weight and the possibility of increased use of alternative fuel vehicles.

includes oil and maintenance, and can also include fixed costs such as insurance, registration fees, financing and the vehicle purchase price. This distinction between marginal and incremental costs is especially important with the growth of services such as Zipcar, Uber and Lyft that offer mobility without the need for vehicle ownership. These services must recover the full average costs of driving per trip and, hence charge larger marginal costs per miles than privately owned vehicles including CAVs.

The Rand Corporation uses data from AAA to estimate that household could save about $5,700 in fixed annual costs by joining a car-sharing program rather than owning a vehicle [10]. They note that these same underlying costs passed back to members in the form of higher per mile rates apportioned over 10,000 miles would be about 57 cents per mile. Added together with the 21-cents-per-mile cost for fuel, maintenance, and tires, the per-mile cost of a car-sharing plan would then be about 77 cents per mile. This compares to about $11–14/h for a Zipcar car in the greater Boston.

Small and Verhoef [11] tally the costs of a typical urban commuting trip in the United States, finding that travel time and reliability—travel time costs (TTC)—together account for 45 % of the average social variable cost, compared to vehicle capital costs (19 %), vehicle operating costs (16 %), and accident costs (16 %).

As seen in Fig. 1, from an individual's cost perspective the total cost per mile $1.02 for new sedan is divided into ownership costs of $0.41/mile (average annual depreciation, license and registration fees, cost of insurance and finance charges)

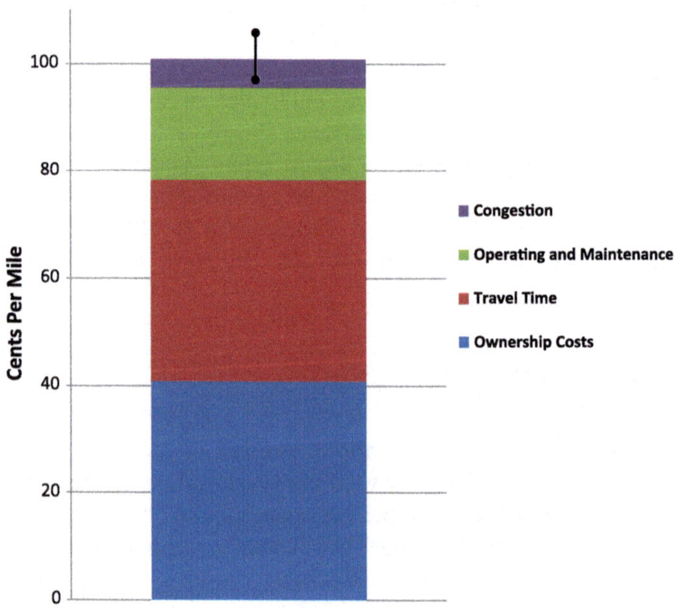

Fig. 1 Average individual cost of travel per mile (cents/mile), range on congestion (*sources* [12, 13] and author's calculations)

[12] divided by the annual miles [14]. The travel time costs of $0.37 is the average time spent driving in free flow traffic divided by the average distance driven daily in the US [15] and multiplied by the average US TTC per hour ($18) [13].

Operations cost of $0.17/mile is the cost of tires, gas and maintenance per mile [12]. Congestion is the average annual individual cost of congestion including time premiums, time delays and additional fuel expenditure per mile [13]. Congestion costs are highly variable from 0 (no congestion) to a small urban area (population less than 500,000) the average congestion costs are 5 cents per mile to a major metropolitan area of 11 cents per mile.

The TTC is one of the largest components of costs borne by individuals. The range of TTCs is large reflecting the wide variation in travel purpose such as commuting to work, while working, or for leisure or shopping. The range in TTC also reflects differing values of time due to socio-economic circumstances such as wealth and age.

Travel demand is usually thought of as derived demand from scheduling many activities including work, recreation, shopping and so forth. This view goes back to the time budgets framework postulated by Becker [16] where individuals and households maximize utility by dividing activities between leisure, wage income and travel time.

3.1 Value of Travel Time

The value of travel time is the implicit monetary value (cost) that individuals place on their time while traveling.[2] The US Department of Transportation (USDOT) summarizes the literature by noting three underlying principles [17]: time spent travelling could be spent productively (e.g., working), recreationally (possibly spending money) and time spent travelling may be unpleasant and cause fatigue, hurting productivity. CAVs are likely to impact all three factors. Drivers could make telephone calls, work on a tablet, or browse the internet, as the vehicle is responsible for route finding and other driving operations. This means that highly automated CAVs will cause a significant decline in TTC. The other advantage CAVs have over, say a train, is that they are a private, door-to-door service. The best comparison might be a private vehicle with a hired driver.

The value of time that affects travel decisions likely depends strongly on factors that vary significantly by individual and are difficult to observe. These factors include individual tax rates, ability to use travel time productively, fatigue or displeasure from travel, ability to adjust working hours and times, and the importance of being on schedule for work or appointments [11, 17–22].

[2]The literature variously refers to costs of travel as travel costs, value of travel time and other variations on the opportunity cost of time spent traveling from one activity or place to another. This paper uses TTC for this common concept.

The papers by Small [19] and Small and Verhoef [11] provide a comprehensive review of the travel time literature. They point out that drivers care not only about the amount and value of time per trip, but also the value of reliability, that is how likely is it that a trip of uncertain travel time can be completed within some expected time costs of congestion.

Shires and De Jong [21] conducted a meta-analysis of TTC accounting for the type of data set (stated and revealed preference), type of transportation, wealth of the country (GDP per capita), the age of the data set, region of country (Northern Europe, Southern Europe, Other), length of distance travelled and the variance. They discovered (largely cross-sectional) income elasticity of the TTC of about 0.5 for business travel, 0.7 for commuting and 0.5 for other passenger transport. Long distances lead to higher TTC for commuting and other purposes.

Abrantes and Wardman [21] performed a meta-study on TTC values drawing on 1749 valuations and 229 British studies from the following modes: in vehicle time, free flow time, congested time, walk time, access time, wait time, headway, departure time shift, and late time. They found a GDP elasticity of 0.9, a distance elasticity of 0.16, and that congestion raised TTC by 34 %. Business travelers were found to have a TTC nearly twice that of non-business travelers and commuters to have a TTC 10 % higher than other non-business travelers.

Driving in congested traffic drives up the TTC of travelers in comparison to free-flowing traffic [11, 19, 21, 23]. Steimetz [23] research uses revealed preference data where drivers have the option of purchasing access to a less congested toll road. He finds that driving in congested conditions is valued more negatively than non-congested driving and that a decomposition that shows about 40 % of TTC in congested time periods is risk and accident avoidance effort. This is significant for CAVs (Levels 3 and 4) since one of their major benefits is automated accident avoidance traffic smoothing.

As noted earlier, there is also the related issue of reliability of travel time [24]. Travelers are willing to pay more for less variation in their travel time. This concept is called value travel time reliability (TTR), and plays a significant role in an agent's decision making process especially for those with a fixed work schedule with a penalty for being late. Carrion and Levinson [24] find that like TTC, TTR varies across regions, time of day and length of journey. Connected vehicles, automated or not, should increase the ability to predict travel times, improve routing and reduce travel time reliability costs for individuals. Small [19] notes that improved information provision may also change the spatial and temporal pattern of congestion.

3.2 CAVs and Rebound Driving

As Small [19] notes, it is widely believed that better in-vehicle amenities, mobile communications, and entertainment devices lower the value of time by making travel time less onerous or more productive. However, as he notes, telecommunications appears to be complementary to rather than a substitute for transportation as individuals and businesses have transformed the way people interact. CAVs are like amenities, but at a whole different level since at Level 3 and above, they can provide work and leisure opportunities while driving. This calls into question the classic time budget framework used in transportation demand modeling.

In the standard analysis of rebound driving, induced travel comes from energy savings from increased vehicle efficiency which lowers the cost-per-distance of travel. These lower costs of travel are then passed on to reduce costs for other industries which lead to reduction in prices and increases in profits [25]. With CAVs, however, decreases in travel costs arise from savings in TTC. For non-business travel, this ought to have the same impact as a reduction in fuel costs per mile to encourage additional driving. For business travel, it is not as clear what the impact might be unless the driver is able to do productive work while traveling in which case per mile costs of travel are also lower. Nonetheless the direction of the effect is clear. CAVs will cause more driving to occur and could cause additional low density development as households and firms adjust to lower transportation costs.

From a welfare theoretic perspective Chan and Gillingham [26] note that there is an important difference between efficiency gains that cause externalities that arise from rebound effects on energy use and those that arise from energy service consumption such as congestion. Energy service rebound effects will increase with increased energy efficiency and lower welfare where as rebound energy with efficiency gains may or may not grow depending on the magnitude of rebound effects. However, lower TTC do not in and of themselves lower the energy use per distance and hence the rebound impact will lower welfare from marginal congestion and increase energy use. The other, indirect (substitution) rebound effects that may occur are from additional driving due to substitution away from public transport as the relative cost of CAVs falls.

4 External Costs of Driving and CAVs

Shown in Fig. 2 are the cost per mile to society from driving, about 18 cents. The cost of accidents at 8 cents per mile is the annual expenditure on automobile accidents in the US divided by the total vehicle miles traveled [1]. The cost of congestion is the absolute wages lost to society by delays caused in congestion for a small urban city [27]. The cost of carbon dioxide of 2 cents per mile is the average tonnage of CO_2 emitted by a car per mile [28] times the cost to society of a ton of carbon dioxide [29]. Criteria pollutants cost of 2 cents per mile [30]. Finally, the cost of fuel insecurity of 1 cent per mile is the estimated macroeconomic disruption component in dollars per gallon of the US dependency on imports [31].

A question arises: are the external costs of driving, on a per-mile basis, significantly influenced by automation? If there are no changes to when and where vehicles are used then this is likely correct. On the other hand, if automation increases the prevalence and use of CAVs relative to connected or non-connected

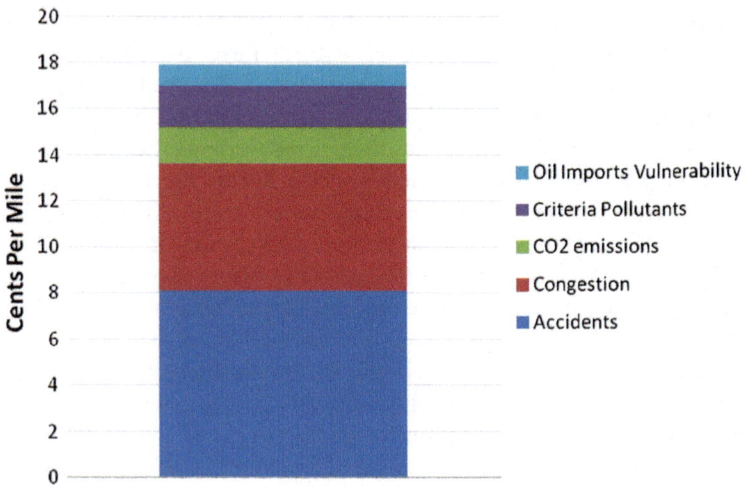

Fig. 2 External costs of driving per mile (2015 cents/mile). *Sources* [1, 13] and author's calculations

conventional vehicles *in situations that have high external costs* then they may have significantly higher external costs per mile. This is likely to be the case for driving in congested areas or times of day. As noted above, one of the advantages of CAVs is that they lower two aspects of TTC with respect to congestion: the value of time costs while in congested traffic and the premium that people are willing to pay to reduce the effort of driving in congested conditions. Both of these effects will, all else equal, make it more desirable and likely to have CAVs driving in congested conditions. This will impose external costs on non-automated vehicles. Advocates of CAVs, to the contrary, often envision a future where all vehicles are automated or in which CAVs drive in special CAV-only lanes.

The experience of high-occupancy-vehicle (HOV) lanes that are common in congested urban areas may be instructive. Their purpose is to encourage more efficient use of highways by increasing the number of passengers travelling per road mile and reducing the number of vehicles in non-HOV lanes. While some HOV lanes have been proven effective, others are found to be underutilized, because carpools account for only a small proportion of total vehicle travel and not all carpoolers use HOV lanes [32]. High occupancy and toll (HOT) lanes allow for single-occupancy vehicles (SOVs) drivers to pay a toll use using HOV lanes. HOT lanes have been adopted by the Los Angeles, San Diego, Houston, Salt Lake City, Denver, and Minneapolis-St Paul metropolitan areas, and many other cities are considering introducing HOT lanes [33]. While HOV lanes can be seen as a sorting mechanism to sort out travelers with different carpool organization costs, HOT lanes can further sort out travelers with different congestion costs [33].

Using a model that ignores the heterogeneity in travelers' congestion costs, Konishi and Mum [33] show that the welfare effects of HOV and HOT improve the social welfare in some cases, but aggravate the situation in others. While HOV lanes encourage carpooling, reducing the total traffic, they can also cause distortion from the difference in congestion levels between HOV and non-HOV. Also, if HOV lanes are converted to HOT lanes this allows single occupancy vehicles in HOV lanes to have an adverse effect by discouraging carpooling. They find, depending on the specific case, that converting HOV lanes to HOT lanes may reduce social welfare. This is likely to also be the case with dedicated CAV lanes. In some cases we will see improvements in traffic volume and flows (platooning, traffic smoothing). In other places, dedicated CAV lanes will make traffic worse for non-automated drivers. Further, despite the optimism for CAVs, non-automated vehicles and CAVs will likely share the roadway for the foreseeable future unless governments are prepared to require that individuals forego active driving.

5 External Costs of Driving and CAVs

The optimal charge for driving requires road users to pay all use and external costs of driving. This includes the costs of road infrastructure, environmental impacts, noise, accidents, and congestion. Parry and Small [22] find that the optimal US gasoline tax is $1.01/gal ($2010) (more than twice the current rate) and that the congestion externality is the largest component of the optimal fuel tax followed by a term that accounts for tax distortions, and then accidents and local air pollutants, with climate change impacts only having a minor component of the tax.

These estimates are not likely to be the same for CAVs since, depending on location, the congestion impacts of CAVs could be larger than convention vehicles. CAVs are likely to have the same per-miles costs with respect to environmental impacts, higher infrastructure costs and lower accident costs. A review of the literature by Lindsey [34] finds that while economists do agree that highway congestion should be solved by pricing, they disagree over how to set tolls, how to cover common costs, what to do with any excess revenues, whether and how "losers" from tolling previously free roads should be compensated, and whether to privatize highways.

In charging for externality costs of congestion there are two perspectives on how to set up an optimal and "fair" system. One way is to simply charge all vehicles the same cost per mile and let vehicle drivers optimize where and when to drive and which type of vehicle to drive. The other perspective is to recognize that CAV drivers face a lower cost per mile of disutility from driving and adjust per mile charges to account for this difference. This latter approach aims to take equity considerations into account: to not let CAVs impose disproportionate costs on conventional vehicles from congestion externalities.

As noted by Parry [22] the framework for designing congestion taxes is well developed, in that we know, in principal, the potentially important factors including network effects, bottlenecks, existing HOV lanes, time of day, marginal external costs and so forth. The best policies will need to be assessed on a case-by-case basis calibrated to local traffic flows, speed-flow relations, and behavioral responses to tolling. Since TTC differs greatly among drivers the best pricing scheme is not a uniform toll across all lanes, but differentiated tolls that allow drivers to sort themselves into more or less congested lanes.

Raising gasoline taxes to account for their external impacts is politically difficult. Mileage charges are increasing be seen as a promising alternative to fuel charges [35, 36]. Mileage charges, while raising privacy concerns, maybe much more tractable with connected vehicles since the real time monitoring necessary for better routing will make collecting accurate spatial and time-of-day vehicle use relatively simple.

6 Final Comments

The advent of highly automated vehicles is exciting to those who see the benefits in terms of safety, access by elderly and the handicapped and for those who view the act of driving as tiring or taking away from other activities. Contrary to some of the optimistic literature on CAVs, there may be some significant unintended consequences to their use. The two areas most clearly of concern are additional rebound driving due to the lowering of TTC. This may lead to increases in low density development and longer, but less costly commute times. Perhaps the growing use of mobility services such as Uber and Lyft and will provide insight into these outcomes.

A second area of concern is the impact of CAVs on congestion in areas of mixed traffic where CAVs operate with non-automated vehicles or pedestrians and bicyclists. There is the potential for some routes to get more congested (because the cost of congestion is less to drivers of CAVs) which will impose costs on non-automated drivers. If automation increases the prevalence and use of CAVs relative to connected or non-connected conventional vehicles in situations that have high external costs then they may have significantly higher external costs per mile. This is likely to be the case for driving in congested areas or times of day. Alternatively, in some areas, superior route and traffic flows from connectivity and synchronized driving may lower congestion and increase route capacity.

References

1. NHTSA (2015) Traffic safety facts, 2013 Data. NHTSA. http://www-nrd.nhtsa.dot.gov/Pubs/812169.pdf
2. NHTSA (2013) Preliminary statement of policy concerning automated vehicles, no. Monograph. http://www.nhtsa.gov/staticfiles/rulemaking/pdf/Automated_Vehicles_Policy.pdf
3. SAE International, International (2014) Taxonomy and definitions for terms related to on-road motor vehicle automated driving systems, vol J3601. http://standards.sae.org/j3016_201401/

4. Brown A, Gonder J, Repac B (2014) An analysis of possible energy impacts of automated vehicle. In: G. Meyer, S. Beiker (eds) Road vehicle automation. Lecture Notes in Mobility. Springer International Publishing, 137–153. http://link.springer.com/chapter/10.1007/978-3-319-05990-7_13
5. MacKenzie D, Wadud Z, Leiby PN (2014) A first order estimate of energy impacts of automated vehicles in the united states. In: Transportation research board annual meeting, Vol. 93rd. The National Academies, Washington, DC
6. Morrow III, William R, Jeffery BG, Sturges A, Saxena S, Gopal A, Millstein D, Shah N, Gilmore EA (2014) Key factors influencing autonomous vehicles' energy and environmental outcome. In: G. Meyer, S. Beiker (eds) Road vehicle automation. Lecture Notes in Mobility, Springer International Publishing, 127–35. http://link.springer.com/chapter/10.1007/978-3-319-05990-7_12
7. Barth M, Boriboonsomsin K, Wu G (2014) Vehicle automation and its potential impacts on energy and emissions. In: G. Meyer, S. Beiker (eds) Road vehicle automation. Lecture Notes in Mobility, Springer International Publishing, pp 103–112
8. Fagnant DJ, Kockelman KM (2014) The travel and environmental implications of shared autonomous vehicles, using agent-based model scenarios. Trans Res Part C: Emerg Technol 40(March):1–13. doi:10.1016/j.trc.2013.12.001
9. Schipper L (2002) Sustainable urban transport in the 21st Century: a new agenda. Transp Res Rec: J Transp Res Board 1792(1)(J Art):12–19
10. Anderson JM, Kalra N, Stanley KD, Sorensen P, Samaras C, Oluwatola O (2014) Autonomous vehicle technology: a guide for policymakers. Rand Corporation, Santa Monica, CA. http://www.rand.org/pubs/research_reports/RR443-1
11. Small KA, Verhoef ET (2007) The economics of urban transportation. Book, Whole. Routledge, New York
12. AAA (2015) Your driving costs: how much are you really paying to drive? http://exchange.aaa.com/wp-content/uploads/2015/04/Your-Driving-Costs-2015.pdf
13. Schrank D, Eisele B, Lomax T, Bak J (2015) "2015 Urban mobility scorecard. Texas A&M Transportation Institute. http://d2dtl5nnlpfr0r.cloudfront.net/tti.tamu.edu/documents/mobility-scorecard-2015.pdf
14. FHWA (2015) Our nation's highways—2000, selected facts and figures. Office of Highway Policy Information Federal Highway Administration. https://www.fhwa.dot.gov/ohim/onh00/bar8.htm
15. Triplett T, Robert S, Sandra R (2015) American driving survey: methodology and year 1 results, May 2013–May 2014. AAA Found Traffic Saf. https://www.aaafoundation.org/sites/default/files/2015AmericanDrivingSurveyRerport.pdf
16. Becker GS (1965) A theory of the allocation of time. Econ J 75(299):493–517. doi:10.2307/2228949
17. Belenky P (2011) Revised departmental guidance on valuation of travel time in economic analysis. US Department of Transportation, Washington, DC
18. Hensher D (2013) Value of travel time savings. In: Andre de P, Robin L, Emile Q, Vickerman (eds) A handbook of transport economics. Edward Elgar Pub, Cheltenham, 135–159
19. Small KA (2012) Valuation of travel time. Econ Transp 1(1–2). doi:10.1016/j.ecotra.2012.09.002
20. Abrantes PAL, Wardman MR (2011) Meta-Analysis of UK values of travel time: an update. Transp Res A: Policy Pract 45(1):1–17. doi:10.1016/j.tra.2010.08.003
21. Shires JD, de Jong GC (2009) An international meta-analysis of values of travel time savings. Eval Program Plann Evaluating Impact Transp Projects: Lessons Disciplines 32(4):315–325. doi:10.1016/j.evalprogplan.2009.06.010
22. Parry IWH, Small KA (2002) Does Britain or the United States have the right gasoline tax? Resources for the Future
23. Steimetz SSC (2008) Defensive driving and the external costs of accidents and travel delays. Transp Res B: Methodol 42(9):703–724. doi:10.1016/j.trb.2008.01.007

24. Carrion C, Levinson D (2012) Value of travel time reliability: a review of current evidence. Transp Res A: Policy Pract 46(4):720–741. doi:10.1016/j.tra.2012.01.003
25. Barker T, Jonathan R (2007) Macroeconomic effects of climate policies on road transport: efficiency agreements versus fuel taxation for the United Kingdom, 2000–2010. Trans Res Rec: J Transp Res Board 2017(-1):54–60
26. Chan NW, Gillingham K (2015) The microeconomic theory of the rebound effect and its welfare implications. J Assoc Environ Res Econ 2(1):133–159. doi:10.1086/680256
27. Schrank D, Eisele B, Lomax T, Bak J (2015) 2015 Urban mobility scorecard. Texas A&M Transportation Institute. http://d2dtl5nnlpfr0r.cloudfront.net/tti.tamu.edu/documents/mobility-scorecard-2015.pdf
28. EPA (2008) Average annual emissions and fuel consumption for gasoline-fueled passenger cars and light trucks. US Environ Prot Agency Off Transp Air Qual. http://www3.epa.gov/otaq/consumer/420f08024.pdf
29. Interagency Working Group on Social Cost of Carbon, United States Government (2013) Technical support document: technical update of the social cost of carbon for regulatory impact analysis—Under executive order 12866. The White House. https://www.whitehouse.gov/sites/default/files/omb/inforeg/social_cost_of_carbon_for_ria_2013_update.pdf
30. EPA, and USDOT (2012) Final rulemaking for 2017–2025 light-duty vehicle greenhouse gas emission standards and corporate average fuel economy standards. http://www3.epa.gov/otaq/climate/documents/420r12901.pdf
31. NHTSA (2012) 2017 and later model year light-duty vehicle greenhouse gas emissions and corporate average fuel economy standards. 40 CFR Parts 85, 86 and 600
32. Li J (2001) Explaining high-occupancy-toll lane use. Transp Res D: Transp Environ 6(1):61–74. doi:10.1016/S1361-9209(00)00013-4
33. Konishi H, Mun S (2010) Carpooling and congestion pricing: HOV and HOT Lanes. Reg Sci Urban Econ 40(4):173–186. doi:10.1016/j.regsciurbeco.2010.03.009
34. Lindsey R (2006) Do economists reach a conclusion on road pricing? The intellectual history of an idea econ journal watch. Econ J Watch 3(J Art):292–379
35. Lindsey R (2010) Reforming road user charges: a research challenge for regional science. J Reg Sci 50(1):471–492. doi:10.1111/j.1467-9787.2009.00639.x
36. Sorensen P, Ecola L, Wachs M (2013) Emerging strategies in mileage-based user fees. Transp Res Rec: J Transp Res Board 2345(J Art):31–38. doi:10.3141/2345-05

The Socio-Economic Impact of Urban Road Automation Scenarios: CityMobil2 Participatory Appraisal Exercise

Carlo Sessa, Adriano Alessandrini, Maxime Flament, Suzanne Hoadley, Francesca Pietroni and Daniele Stam

Abstract This document aims at assessing and fine tuning alternative scenarios concerning road automated transport, based on the contribution of research, industry and public stakeholders convened at the CityMobil2 Workshops organised in La Rochelle on 30–31st March 2015. Two different paradigms—with and without a shift to shared mobility—were debated and a number of potential socio-economic impacts were identified. Road automation scenarios are devised for different urban typologies—large metropolitan areas, polycentric city networks, small-medium towns, rural/tourist areas. Impacts are assessed in a qualitative fashion—with the support of an online DELPHI survey followed by the workshop debates—in relation to a number of variables. These include: job disruption and creation; personal trips costs; public budget effects; insurance costs; accessibility to remote areas; road capacity and its use; journey comfort and convenience; energy and emissions; land saving for new public space uses; social impacts in terms of

C. Sessa (✉) · F. Pietroni
ISIS - Institute of Studies for the Integration of Systems, Largo dei Lombardi 4, Rome, Italy
e-mail: csessa@isinnova.org

F. Pietroni
e-mail: fpietroni@isinnova.org

A. Alessandrini · D. Stam
CTL – Centre for Transport and Logistics Sapienza, University of Rome, Via Eudossiana 18, Rome, Italy
e-mail: adriano.alessandrini@uniroma1.it

D. Stam
e-mail: daniele.stam@uniroma1.it

M. Flament
ERTICO ITS Europe, Avenue Louise 326, Brussels, Belgium
e-mail: m.flament@mail.ertico.com

S. Hoadley
POLIS – European Cities and Regions Networking for Innovative Transport Solutions, Rue du Trône 98, Brussels, Belgium
e-mail: shoadley@polisnetwork.eu

G. Meyer and S. Beiker (eds.), *Road Vehicle Automation 3*,
Lecture Notes in Mobility, DOI 10.1007/978-3-319-40503-2_13

safety, personal security, health and active travel (trade-offs in automated rides vs. walking or cycling) and different perception/value of time spent travelling in automated vehicles.

Keywords Automated transport · Collective · Private · Scenario · Modal share · Car ownership · Survey · Urban sprawl · Compact city · City network · Private car use · Shared transport · Public transport · Walking and cycling · Future · Challenges

1 Introduction

On 30–31st March 2015, more than 100 experts from Europe, the US, Japan and Singapore met in La Rochelle (France) in a workshop organized by the European project CityMobil2. The workshop focused on the expected impacts of road vehicle automation take up in different typologies of urban environments—compact cities, sprawled cities, connected cities and rural areas—and for two different scenarios: automation of private cars and diffusion of shared self-driving vehicles. The experts had to assess the potential impacts of automation on the economy, transport, the environment and society.

The pros and cons of two "caricature" scenarios—automation with and without a paradigm shift to shared mobility—were debated and a number of potential impacts were identified in terms of: job disruption and creation; personal trips costs; public budget effects; insurance costs; accessibility to remote areas; road capacity and its use; journey comfort and convenience; energy and emissions; land saving for new public space uses; social impacts in terms of safety, personal security, health and active travel (trade-offs in automated rides vs. walking or cycling) and different perception/value of time spent travelling in automated vehicles.

Preliminary analyses undertaken to prepare for the workshop included a review of recent urban self-driving transport scenario studies, an online DELPHI survey, and a qualitative evaluation of the socio-economic impacts of different urban road automated transport scenarios.

In this paper we summarize the key results of:

- The online survey focusing on road transport automation in different urban contexts.
- The qualitative appraisal of the expected impacts of driverless urban transport scenarios and the results of the 1st Day session discussing the impacts.
- The 2nd Day session on the stakeholders' perspectives concerning the preferred scenario, and which business model changes and policies would be needed to enable the transition to the preferred scenario of urban transport automation.

2 Results of the Online Survey

The aim of the on-line survey was to evaluate 8 options of urban transport automation, contrasting 2 extreme scenarios (automated private car ownership vs. automated car-fleet sharing) in four different urban typologies:

- *Urban Sprawl* Large cities with a city core surrounded by low density suburbs, with the prevalence of fast trips mostly done alone to/from the city centre and—to a limited extent—of tangential trips. Car ownership is high, the daily trips per capita in a working day are high, the average distance is high and the occupancy rate is low.
- *City Network* Polycentric regions/city networks with the prevalence of fast trips mostly done together. Car ownership is low, while all other benchmark values—daily trips per capita, average distance occupancy rate—are high.
- *Small Compact City* The prevalence of short distance/slow trips, done together or by foot and bicycle, characterize small compact cities. The occupancy rate is high, while all the other benchmark values—car ownership, daily trips per capita, average distance—are low.
- *Rural/Tourist Areas* Low density areas with the prevalence of slow trips mostly done alone. Car ownership is low, the daily trips per capita are low, the average distance is high and the occupancy rate is low.

A key assumption underpinning this approach is that—besides some general technological and social drivers—urban transport automation challenges, opportunities and impacts will be different in low and high density city contexts, and depending on the available transport infrastructure (in particular the existence of high capacity links). The survey was concentrated only on the direct impact on transport patterns in each urban form, not on second order land use impacts of automation on the urban forms themselves—for instance the extent to which automation may provide a further impulse to urban sprawl facilitating longer journeys is beyond the scope of our study.

For each urban form, respondents to the online survey had to consider two contrasted and somehow “extreme” scenarios:

- *Scenario 1: Automated car ownership-centred mobility.* A private automated mobility scenario is the result of a technology revolution without a significant change in the conventional private transport behaviour. Most of the people will continue to own and drive their cars. Self-driving vehicle sharing will develop to a limited extent, within the same household—reducing in some contexts the need to purchase a second car—or more broadly by means of peer-to-peer car sharing schemes. In any event, the autonomous vehicles will continue to be mostly in private ownership.
- *Scenario 2: Automated car fleets-centred mobility.* This scenario envisages a shift from privately owned individual vehicles to collective purchase and operation of fleets of self-driving vehicles—that may be owned by private or

public service operators—and are available for simultaneous (ride-sharing) or sequential use (car sharing) on demand, complementing and integrating existing mass transits.

89 participants answered to the survey. The majority of respondents was interested to evaluate automation scenarios only in the urban sprawl context, considering the other contexts less relevant or because they were less confident in assessing likely self-driving scenarios for other urban forms. The car ownership centered scenarios was also slightly more popular than the shared self-driving vehicles, collecting more answers. Figure 1 summarizes the main results concerning the expected changes of four key variables characterizing urban mobility—daily trips per capita, average journey distance, occupancy rate and car ownership—for the 8 options considered (2 scenarios × 4 urban forms):

The arrows (red for scenario 1 and yellow for scenario 2) represent for each key variable both the direction (increase, decrease or stability) and the intensity (bigger arrows for >30 % change in the base variable, smaller arrows for a change between 10 and 30 %) of the likely change, according to the most frequent responses to the survey (modal values).

The most frequent answers have been more conservative than our impact assumptions presented in the online questionnaire. According to the majority of respondents urban transport automation will cause the key variables to change within the range 10–30 % at most—or to stay the same—not changing radically (more than 30 %) in one direction or the other. This is because—in the opinion of many—autonomous vehicles, are only one of many factors that will affect transport demands and costs in the next few decades, and not necessarily the most important. More in detail, the key insights and conclusions for the single variables characterizing urban mobility are as follows:

- *Daily trips per capita* will increase in the urban sprawl and rural areas settings, as the self-driving car availability will augment the flexibility and opportunity to combine daily travel schedules for different members of the household. In the more compact forms—city network and small compact city—daily trips are expected to increase only in the automated car-fleet scenario, thanks to the availability of more capillary services. The impact pathway presented in the survey assumes that the cars are more often available because of their capability of self-driving, and this alone will induce more daily trips per capita (increasing more than 30 %). Most of the respondents to the survey were more prudent, guessing for a more moderate increase, as car availability is not the only factor affecting car use, especially in potentially congested urban contexts or where a good high capacity public transport is available. The number of impaired mobility people trips should increase, but it also depends on the availability of the facilities at the end of the journey and on the easiness in getting in and out of a car. By the same token, the impact on aged people propensity to travel may be important, but insofar as the aging population do not easily understand and adopt new technologies may be scared by these developments.

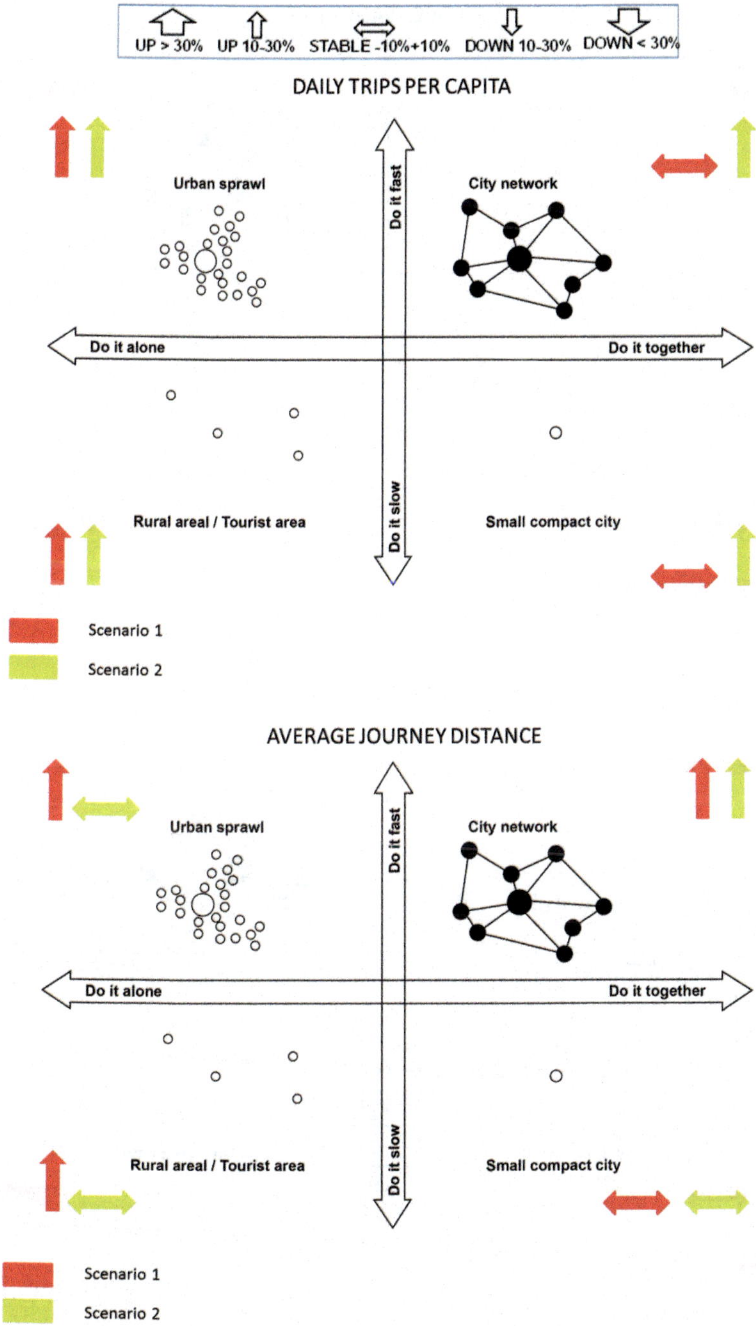

Fig. 1 Survey results

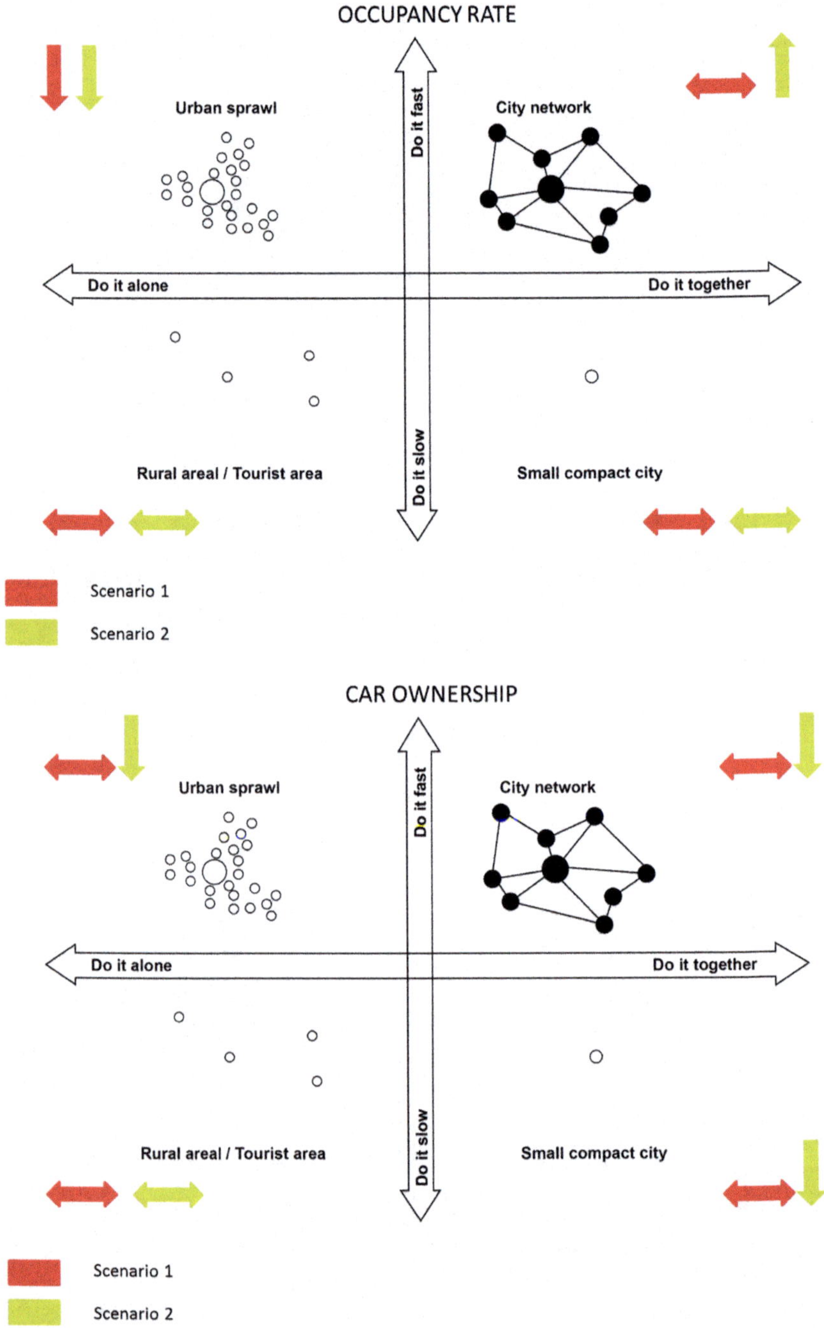

Fig. 1 (continued)

- The *average journey distance* will increase in the private automated scenarios for all urban forms, except in the small compact city, where short distance trips are prevailing and self-driving will not change substantially the range of accessibility choices. On the contrary, the average journey distance will not increase in all car-fleet automated scenarios, except in the city network, where the offer of coordinated car sharing and ride sharing options is likely to increase the longer trips between the different cities of the network. The impact pathway presented in the survey assumes that the car use for longer trips is encouraged because the trips become more comfortable and the passengers are free to choose what to do while the car is driving itself. Average distance may increase between 10 and 30 % as a result. Most of the respondents to the survey agreed on this assumption. However, a consistent minority were more skeptical due to the higher autonomous vehicle costs, which may reduce both the penetration of these vehicles in the market and their extensive operation and use by households members. In addition, the length of the trip is primarily affected by the current city size and form, a factor that influences travel needs and cannot be changed in short times. Both factors—low driverless cars penetration and rigid land use patterns—may cause average distance not to increase, at least in the short term. The average commuting time may also remain constant, as automated modes will not automatically be faster—indeed speed limitations for the autonomous driving are in the cards. The picture can obviously change in the long term, as the greater travel comfort can induce further urban sprawl and longer commuting trips.
- The *occupancy rate* will decrease in the urban sprawl context, as an effect of the empty trips to relocate the self-driving cars to the next users—i.e. another member of the household in the private automated scenario or another user in the car-fleet scenario. This effect is not considered significant instead in other urban contexts (small compact cities, rural/tourist areas), with the exception of the car-fleet scenario in the city network, where fleet based car sharing and ride sharing services are assumed to optimize the journeys and bring an increased occupancy rate (between 10 and 30 % more). The impact pathway presented in the survey assumes that empty trips will increase substantially (causing an average occupancy rate increase of 30 % or more) in the private automated mobility scenario, to allow different members of the household to use the same car during different hours of the day. Automation will not deliver the same effect in the car fleet scenarios, because fleet owners will be motivated to minimize empty running, e.g. through dynamic pricing. Most of the respondents to the survey consider the assumption for the private automated mobility scenario too pessimistic. Occupancy rates—some respondents claimed—are already low especially in the urban sprawl context (around 1.3), it is difficult to reduce them further. In addition, the operating costs of "dead-heading" empty private vehicles will become something households examine, pushing for a more efficient use of the car. Empty trips could be reduced as well by the sharing of self-driving cars between members of the same household or trough ride sharing with neighbors or work colleagues. In a nutshell, a decrease of occupancy rate is

expected, but more moderate—below 30 %—in the private automated mobility scenarios, and not expected at all in the automated car fleet scenarios. In the city network automated car fleet scenario—as mentioned—the occupancy rate will increase, as most of the respondents to the survey agree.

- Finally, according to the majority of respondents, *car ownership* is poorly affected in the private automation scenario—whatever the urban form. On the contrary, it is obviously likely to decrease in the car-fleet sharing scenarios, but the latter not in the rural area context, where the car will remain a key asset to hold (with more opportunities however for ride sharing or peer-to-peer sharing). However, some respondents to the survey highlight that car ownership could decrease substantially also in the private automated mobility scenarios, because self-driving cars may serve the mobility need of more than one family member in the same day, and the ownership of second or third cars could drop for this reason. If the autonomous vehicles are more expensive than the conventional ones, new vehicles purchase will be also limited, with a detrimental effect on car ownership. On the other hand, in the automated car fleet based scenario—according to some respondents car ownership will decline if car and/or ride sharing will effectively happen—in particular in the city network context where a radical decrease of car ownership is assumed (more than 30 % of decrease). However, for this to happen it will require new business to start up, which will need payback to cover for car purchase, depreciation, maintenance, insurance and fuel, and based in the right places. Thus, the cost could be high to the consumer, which may mean that the adoption takes longer and thus car ownership would not change so rapidly.

As it concerns the changes of modal share, between *private car use, shared transport, public transport and walking and cycling*, the results of the survey are shown for the two scenarios in the Figs. 2, 3, 4 and 5:

Not surprisingly, in the private automated scenario the private car use is expected to increase for all urban contexts, as a consequence of the greater comfort of using and travelling with a self-driving car. The only exception is observed in the rural area context, where the majority of respondents think private car use will remain the same. A reduction of the public transport share is expected, almost mirroring the increase of the private car use, while most of the respondents think that walking and cycling shares will remain stable, as automated transport should not attract those that enjoy walking or cycling.

The impact pathway presented in the survey assumes for modal shares in the private automated scenario an increase of private car use. This primarily because thanks to new self-driving capabilities cars can be used by people that cannot drive (children and elderly people with reduced driving capabilities). In addition, the pathway assumes stable share of car/ride sharing especially in the urban sprawl context, a reduction of public transport ridership, and a slight negative trend from walking and cycling, mainly due to the trips made with new automated vehicles when the conditions for walking or cycling are not comfortable (e.g. bad weather). Further comments from the respondents to the survey pointed towards a possible

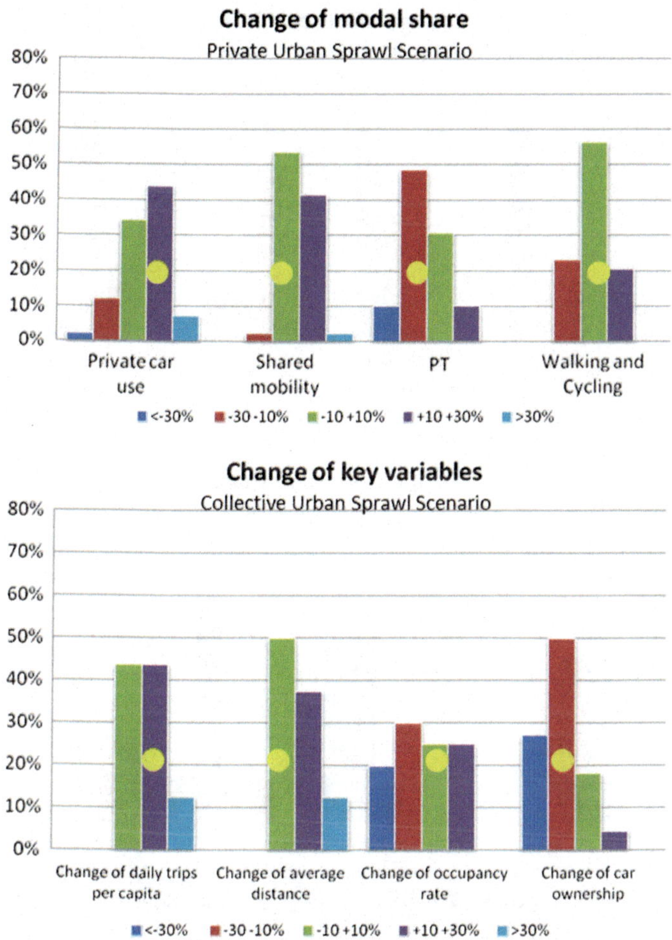

Fig. 2 Survey results for the model share scenario—Urban Sprawl

increase of shared transport also in this scenario. Shared mobility will be higher if cars become available to younger people who currently travel by public transport, and the acquisition of private—and expensive—automated vehicles will probably encourage their owners to propose more ride sharing to others to amortize the purchase costs. Some peer-to-peer car sharing will be also encouraged—although less than ride sharing—as connected and automated features of the new cars will reassure owners and let them share their cars more easily, reducing the risks of accidents, thefts, etc., and ensuring that the cars come back to the owners when needed. Finally, some respondents questioned the expected reduction of the public transport share. This depends by what will happen with the costs of the different options for the user: self-driving, shared transport, public transport. Insofar as the prices of automated vehicles will be higher, this will reduce private car usage by

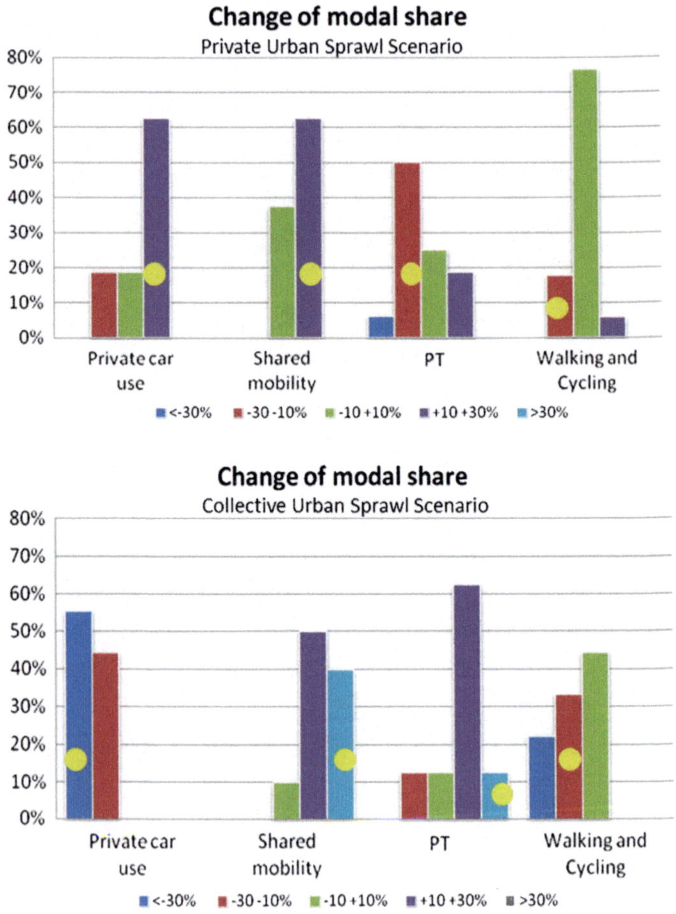

Fig. 3 Survey results for the model share scenario—City Network

raising public transport and shared mobility. In addition, if the circulation of self-driving private vehicles in the urban areas will be more easily controlled and managed, this may have a positive effect also on the reliability of public transport in the same areas, increasing its use. In the car-fleet centered scenario, there is a potential complementarity between public transport and self-driving shared transport modes in the city network context, as both shares are expected to increase while private car use is likely to decrease radically. The same effect is not expected in the small compact city and in the rural area contexts, where shared transport will increase but the public transport share is likely to remain the same. On the contrary, in the urban sprawl context the new self-driving shared transport mode can be a potential substitute for public transport, with shared mobility eroding not only the private car use but also the public transport share. The impact pathway presented in the survey assumes for modal shares in the automated car fleet based scenario that

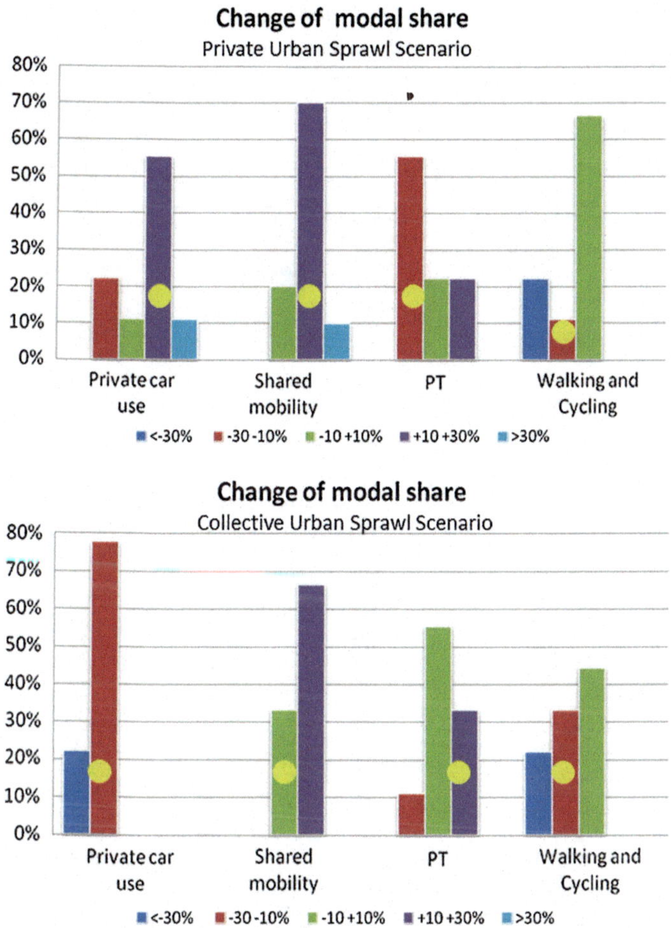

Fig. 4 Survey results for the model share scenario—Small Compact City

the car use decreases because efficient public transport becomes available where it was not before the automation (last-mile public transport), increasing the share of seamless public transport intermodal trips. Shared mobility is also increased a lot as the availability of fleets of shareable self-driving cars is the main feature of this scenario, while soft modes are not affected. Most of the respondents agreed with these assumptions. Although not mentioned in the scenario, driverless taxis will be a form of shared mobility, and they will increase substantially. Public transport might see even a rise of high capacity arterials (e.g. metro rides) since publicly run and maintained automated vehicles might serve as feeders thus offering for the first time especially in sprawled areas—a competitive public transport option. However, one comment "out of the chore" highlighted that shared mobility services would not hold the same characteristics (e.g. response time) in central/high demand

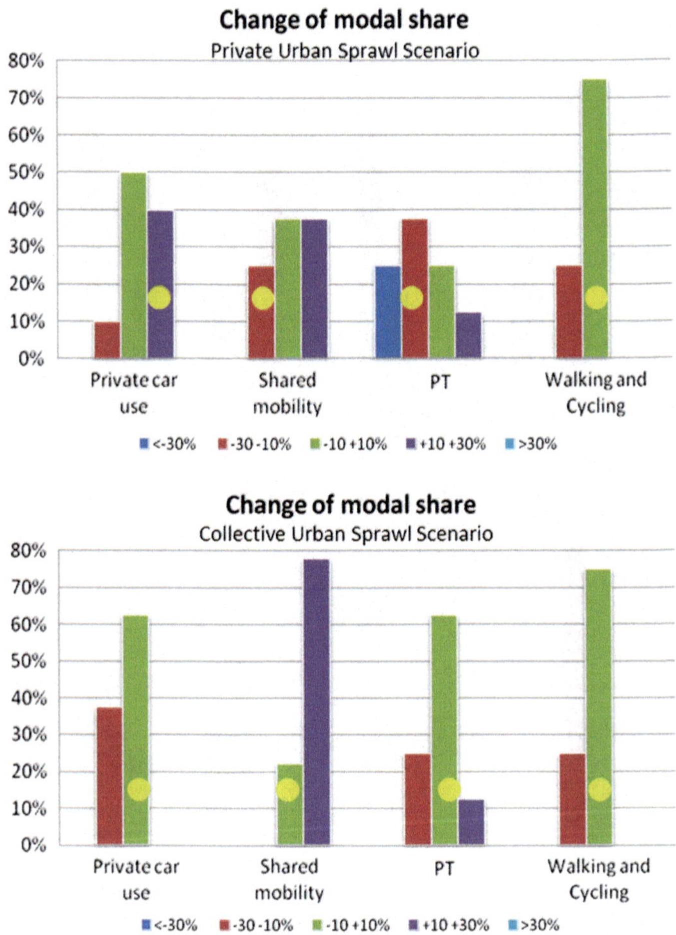

Fig. 5 Survey results for the model share scenario—Rural/Touristic Area

and in peripheral/low demand areas, and the same applies to conventional public transport services. Although it is true indeed that shared vehicles could offer a solution for the last mile problem, this would not dramatically change the level of service between central and peripheral zones, and the households living in peripheral locations might choose to still own and use a private vehicle. Finally, a potential positive side-effect on walking and cycling has been also mentioned, as the new free space due to less need of parking space for the self-driving cars (which are expected to circulate more continuously during their lifetime) may lead to reconversion of parking lot space to more attractive pedestrian zones. This means that more people might prefer to walk due to enhanced safety, walking space and less pollution.

3 Qualitative Appraisal of Expected Impacts and Highlights from the 1s Day Session

The number of trips per capita, the average travelled distance and the occupancy rate of each transport mode are the key variables to determine the number of vehicle-kilometers travelled each day. This, together with the modal share of the different modes, allows to know whether in each scenario the number of vehicle-kilometer travelled overall increases or decreases with respect to the "do nothing" scenario and how much. Most of society and environment impacts of transport depends on the vehicle-kilometer travelled. Even if an automated vehicle can be less polluting or less prone to accidents than a manually driven one, the overall impact might still be negative if the increase in the number of vehicle-kilometer (exposure) is more than the reduction of accident risk or emission per vehicle-kilometer, producing a rebound effect. Similarly, the economic impacts are dependent on the car ownership rate and the vehicle-kilometers travelled because these variables influence the number of vehicles sold and the economy related to fleet maintenance and management. The key variables considered in the online survey have then be used to give a qualitative evaluation of 13 long-term socio-economic impacts belonging to 4 evaluation categories on the basis of the survey indicator results, as illustrated in the scheme below (Fig. 6).

A first qualitative appraisal of impacts has been presented at the La Rochelle workshop. The results of the discussions are summarized below for the four categories of impacts.

3.1 Economy

The *economic impacts* computed with the qualitative methodology included *new jobs, employment, personal trip costs, fines, and the impacts on insurance costs and*

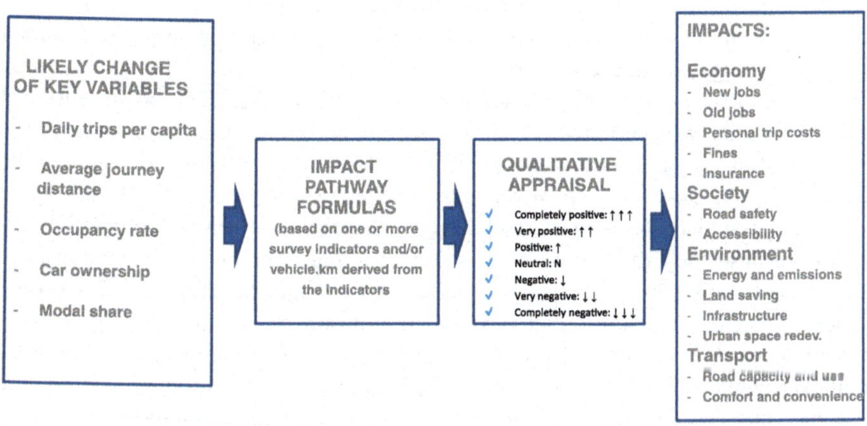

Fig. 6 Evaluation of survey results

services. According to the computations, all economic impacts will be positive in the private automated scenario, in particular in the urban sprawl context but also, with slight differences of intensity, in the other urban contexts (city network, small compact city, rural area). This is caused primarily by the significant increase of total mileage expected when autonomous vehicles will be diffused, making travel more comfortable and accessible to categories of users—elderly, disabled—that today are excluded. Impacts will be positive also in the shared transport scenario, with the only exception of employment in old jobs, where the effect is considered neutral, because traditional jobs in the car manufacturing, repair, maintenance etc. will not increase due to the reduction of cars sold on the market. However, according to most of the survey respondents the impact on employment may be less favorable for the private automated scenario, in all urban contexts, because of job losses in maintenance and control services needed per km travelled, not compensated by the increase of total mileage. Moreover, other economic impacts include:

- The impact of travel comfort on personal productivity, during and after the trip
- The impact of safety on human capital health and productive value
- The impact of accessibility enabling economic development, in particular of more remote suburban areas where self-driving cars contribute to improve accessibility
- The impact of fines not only on household budgets, but also on public budgets that will suffer a loss.
- The same for parking fees: their reduction is a benefit in terms of personnel trip costs, but would have an heavy negative impact on the local authorities budget, as parking charges are an important source of revenue

3.2 Society

The *social impacts* computed with the qualitative methodology include *safety and accessibility for disabled and elderly people*. According to the computations, these social impacts will be positive in the collective automated scenario, for all urban contexts, while in the private automated scenario the impact on safety is assumed moderately negative, as the reduction of self-driving vehicle accident risk would be more than offset by a significant increase in the total mileage. The positive impact on accessibility will be higher in the private automated scenario compared to the shared self-driving transport scenario, as in the former privately owned cars will be available at the door-step. However, most of the survey respondents do not agree with the pessimistic forecast of road safety decrease in the private automated scenario, because they think improving safety is a must for introducing automated transport—a new technology cannot succeed if it eventually reduces safety on the roads. Moreover, the substantial increase of exposure to risk in the private automated scenario is considered plausible only for the urban sprawl and rural areas contexts, while the increase of mileage is expected to be less significant in the compact city and city networks contexts. Other social impacts include:

- Health: what impact will automated demand responsive vehicles have on our health? Will we cease to walk or ride a bike? Cities are promoting active travel today, especially for the first/last mile.
- Well-being/quality of life.
- Urban space redesign: with fewer private cars in the city, there would be the opportunity to use parking facilities for other purposes (offices, homes) leading to new high quality urban fabric, which is denser without giving the impression of higher density.
- Residential relocation: on the one hand, automation may offer the option of moving away from the city to areas where housing is cheaper. On the other, it may induce forced relocation because accessible areas in the city could push up property prices, thereby pushing poorer people out of the city.
- Improved access to employment—the absence of transport is no longer a barrier, unless it is unaffordable.
- The perception of travel time will change—as it will be possible to work or sleep while travelling.

3.3 Environment

The *environmental impacts* computed with the qualitative methodology include *energy and emissions, land saving, urban space requalification and infrastructure modification*. According to the computations, energy and emissions impacts will worsen in the private automated scenario, due to the increased mileage not compensated by better vehicle and driving performances. In this scenario, the other environmental impacts are expected instead to remain more or less the same (with the exception of the impact on land saving and urban requalification in the rural/touristic area context, which is expected to worsen). In the shared self-driving transport scenario, the environmental impacts are always expected to improve—with the exception of infrastructure modification—for all four urban contexts, and particularly favorable for land saving and urban requalification in the city network context. However, most of the survey respondents do not agree with the pessimistic forecast of increasing energy and emissions in the private automated scenario whatever the urban context, as they think the total mileage will increase significantly only in the urban sprawl and rural area contexts, not in the other more compact urban contexts (small compact city and city network). The most relevant insights from the workshop discussion of environmental impacts include the following:

- Private automation may increase accessibility of remote areas and facilitate urban sprawl. This will cause an increase in distances and in number of trips and may naturally lead to shift away from environmentally friendly modes such as soft and PT.
- On the contrary, collective scenarios may increase *urbanisation* by attracting citizens to live where flexible mobility options are available. The city in this

scenario may be manageable *through integrated PT management* in which all available mobility solutions are considered and can be compared in relation with many different criteria such as cost, time, comfort, need to drive/wish to "do something else than driving" or environmental impacts. In any case, it is important to avoid mode shifts from soft and collective modes (and understand how to do this).

- Concerning *energy and emissions* in the private scenario the main concern in the urban sprawl pattern is the increased mileage that cannot be compensated by better vehicle performances, use of platoons and lower cruising speeds. Only increased use of carpooling solutions may compensate the increased VMT. For vehicles able to search parking on their own, the vehicle owner may be more inclined to enter city centres without the burden to search for a parking place or pay for one as the vehicle may even drive a couple of km to reach an empty parking place. This is highly unattractive for city authorities.
- In the collective scenario the *investment in fleets* offers eventually much more potential to further improve the system gradually as demand increases. Most waste comes from relocation of the empty vehicles but this may be balanced by the increased carpooling which becomes a kind of flexible public transport system.
- In the private scenario most of the *land saving* is connected with parking in urban areas. In the urban sprawl case there will be a high demand on land use outside city areas and higher infrastructure and city running costs. In the collective scenario there will be positive impacts in land use due to a very low need for car parks and no need for car parks in city centres. It will be easier to manage the interface with the public transport and car sharing fleets.
- A private scenario will have less *impact on infrastructure modifications* in comparison with the collective one. On a macroscopic point of view, the *impacts on urban requalification* will be negligible in terms of road network length because there will be the need to maintain an urban road network which can accommodate for both automated and manual vehicles. However, in a collective scenario, part of the urban environment is converted to full automation mixed with pedestrian and cyclist traffic. This may lead to a radical requalification of the centre urban environment, offering to cities the option to become *more liveable* keeping vehicles outside the city centres. It will give the opportunity to rethink the urban environment for pedestrians, autonomous vehicles and deliveries. The need of dedicated lanes will require investments in *infrastructure modifications*.

3.4 Transport

The *transport impacts* computed with the qualitative methodology include *road capacity and use, and travel comfort/convenience*. According to the computations, for the private automated scenario these impacts are expected to be respectively neutral (road capacity) and highly positive (comfort) in all the urban contexts. The road capacity constraints are lessened in the shared self-driving transport scenario, as the total expected mileage is lower than in the private scenario, while the comfort of sharing is considered still positive, but more moderately. However, also in the private scenario the road capacity constraints are somehow lessened in the compact and especially in the city network contexts, because the increase of total mileage is lower than in the urban sprawl and rural contexts. Finally, the impact on road capacity will be particularly favorable for rural/touristic areas in the shared self-driving transport scenario. Besides road capacity and travel comfort, the participants wanted to re-discuss some of the key variables in the survey, to suggest few other elements for the qualitative impact evaluation procedure:

- Modal share is expected not to change in the urban sprawl context for the private automated mobility scenario, while to significantly favour public transport in the collective scenario (with shared automated vehicles mostly serving last mile legs of high capacity public transport routes).
- Earlier adopters of automation would benefit the greatest. Too many automated vehicles would be detrimental to mobility due to limited space. Automated small vehicles (whether private or collective/shared) cannot replace the high capacity public transport systems (bus, tram, etc.). This capacity issue may be resolved with pods operating in 'train mode'.
- The overall travelled vehicle-kilometre are expected to increase significantly (very negative impact ↓↓) in the urban sprawl context for the private automated mobility scenario and even worse (↓↓↓) for the collective scenario (due to self-driving empty trips)
- Few more impacts were asked to be considered, including *travel time*, its *reliability* and the *connectivity* (maybe overlapping with the accessibility in the social category). Travel time is expected to be negatively affected by congestion in the private automated mobility scenario and be very positive in the collective scenario, while its reliability would require further investigation. Connectivity is expected to be positive in both scenarios for the urban sprawl context.

4 Assessment of Survey Results by Stakeholders

4.1 Public Authorities

The *urban transport automation scenarios*—private cars automation versus shared self-driving—are two extreme pictures of the future, real future developments will depend on the environment and what the public authority and the users want. However, it is likely that the automated car ownership-centered mobility scenario will drive this domain forward because the private sector has the means to drive technology development. The public sector is not in a position to make such large investments. The real scenario likely to emerge will also depend on the value of space and peoples' attitude towards car ownership. The value of space is far higher in city centers and compact cities than in the rural environment. To achieve successful automation, there is in any event the need to engage with other stakeholder groups, such as the freight sector and vehicle manufacturers. Diverging views between the public sector and Original Equipment Manufacturers (OEMs) is an obstacle to engagement, i.e., the OEMs business is to sell cars whereas city authorities are striving to reduce car movements. There is therefore a higher potential for collaboration between service providers (private bus operators, car service providers) and public authorities. Freight should not be overlooked too. Car-borne shopping can be replaced by internet shopping and home deliveries. Ideally, vehicles should be adaptable so that they can carry both people and goods.

About the needed *business model changes*, some key issues and insights are apparent:

- *Public or private business?* The key issue for shared services is to figure out the dominant business model, and if this should be publicly or privately run. There is a widely held view that private sector actors will have to take the risk. However, cities need to support them. Cities need to follow and anticipate what the private sector is doing, e.g. car-sharing, Uber. Should the city allow that to happen and simply adapt to it? In any event, in the private business model, if there is no profit for the bus/transport operator company, then it will not operate the shared service. One use case offering an interesting business case is a Park and Ride shuttle service to the city centre.
- *Urban automation as a city policy focus?* The majority of city authorities are not talking about automation—currently, the key issues are active modes and eBikes.
- *Scenarios for coping with uncertain future developments?* It is difficult for cities to make decisions when there are so many unknowns. Hence, the importance of building scenarios and understanding the potential and the risk of these types of services.
- *Avoid generalizations?* Each city is very different and may need to think of a business model for each of the applications of automated systems (CM1). For instance, a first/last mile solution may be not relevant in the inner parts of cities like London/Brussels where everybody is close to a public transport stop.

Finally, about the needed policy changes, the following key issues and insights emerged in the discussion:

- There is a lack of joined up thinking within the EC; for instance, automation is not mentioned in the SUMPs discussion.
- Automation can make underused urban space/land (i.e. car parks) available for other productive uses. It also allows the provision of services of different size/speeds and the possibility of public or private operations. What is paramount is the integration of transport services, where cities do have a role to play.
- Cities need to think about street design to allow the penetration of these cars whilst retaining liveable cities. A crucial point for cities is car ownership. New developments are low car intensity. Are the new developments today fit for

purpose for the next 20+ years? Car ownership is not addressed in the European urban mobility package.

- In relation to the issue of active modes (walking and cycling) for the first mile/last mile, a CityMobil analysis found that less than one-quarter of modal shift came from cars, more came from public transport and cycling/walking. It is important to limit such substitution effect, in order to gain previous car users that will find convenient to shift to alternative modes
- The public transport sector views other innovations as competition (e.g. car sharing/clubs). If automation is detrimental to public sector patronage, then it becomes unviable.
- There are diverging views regarding C-ITS as an enabler of automation.

4.2 *Automotive Industry*

About the proposed urban transport automation scenarios, the possibly most plausible solution is a *complementarity of automation scenarios*. Mixed solutions of transport will be necessary in most of the urban contexts considered (urban sprawl, compact cities and city networks, rural areas). Furthermore, reaching ultimately the level 5 of automation is not really a necessity for the industry, and for practical reasons it may not be needed to automate everywhere all the time. Where it makes sense, in city centers, around suburban PT nodes and in industry parks, last mile solutions to PT are demanded and the collective scenario may be the best solution.

If there is a demand in this direction, the industry will invest and provide solutions to the city authorities. Indeed, it will be up to the city authorities to establish clear policies and urban requalification programs to facilitate the introduction and the operation of these vehicles. Unless there is no clear demand and critical mass of political willingness, it is unlikely that the automotive industry will invest. If this status quo will continue, the private scenario may be the most likely. Most investments will go to the improvement of current vehicles and the preservation of the vehicle manufacturer's business model which is the sales of vehicles rather than an integrated mobility service. In this case, vehicle automation will increase gradually starting from high-end vehicles and operate in less and less complicated traffic environment starting from congested highways, then autopilot on highways and gradually on arteries and in intersections. Also the automation of the parking task may be realized first in parking houses then on street. This scenario is the most likely if there is little unified policy to encourage alternate integrated mobility solutions in urban environment.

A key aspect and future driver is how to develop and implement the needed standards and regulations for different typologies of roads and users in the different scenarios. and. A *framework* to authorize or not the automation in given parts of the road network will be highly appreciated by the industry. It would mean that the road operator commits to a given level of maintenance and enforcement on the authorized sections of the road network. Maintenance is meant in terms of road markings, posting, digital maps, traffic and incident information, etc. Enforcement is meant in terms of not authorized parking, speeding, loading/unloading, etc. There is also a need for proper *standards* to ensure good level of reliability in automation of vehicles (something like the "functional safety standards"). Finally, in order to best address the industrial challenges, it is important to create *clusters of competences* and expertise bringing together automotive, robotics, IT, telecom and consumer electronics.

About the needed *business model changes*, the industry, in general, is already used to answer to the highest potential demand of the market. In this respect, it is clear that currently the priority is for the private automation scenario. There are a couple of market-driven needs that are clearly addressed by the industry:

- The avoidance of higher-risk situations related to monotonous and stressful driving tasks. This is driven by customer demands but also under the pressure of public awareness and societal needs.
- The accessibility for ageing population. The baby boomers are the most inclined to keep opt for private vehicles rather than other modes in the next decades. This represents a very high potential market but increased pressure on safety of elderly drivers may deplete this user base. New vehicles put on the market need to consider the needs of elderly drivers and user-centric automation may be part of the answer.

The automotive industry traditional business model may be challenged by threats that are at the same time opportunities for new entrants:

- Shifts to more mobility services with a higher importance of fleets e.g. car sharing
- Gradual downturn of dealerships: Less sales through dealerships; less maintenance; less fender benders and accidents
- Shift of current players along the value chain: OEM start operating fleet where emerging OEMs may even leapfrog others and drive innovation through partnership with IT industry.

So, if automation is gradually increasing, the traditional business model of the automotive industry is put under high pressure. Automation and electrification combined may decrease dramatically the after sales and maintenance incomes. The latter are strategic, as the parts business accounts for only 10–15 % of sales, but it typically averages 25–50 % of profits. Additionally, profit margins on service contracts are around 50 %, with strong impact on customer loyalty. In addition, sales price may be greatly impacted if the increased product liability of the vehicle manufacturers is passed on to the end-users. In case, the lost income of after-sales and maintenance and the increased liability risks directly to affect the vehicle prices, and it is unclear today if the individual end-users would like to go for a private scenario and would be ready to pay for it. Some premium vehicle owners, sales persons and long distance travelers will certainly keep this scenario option, but other options will be considered for the majority of commuters.

At the end of the day, the most likely option may be the “status quo”, meaning the use of manually driven vehicles. But with some incentives and good political will the scenario may shift to the collective one. So there is a great need to try to understand what the users really want now and in the future, and how they will be able to afford it given the impacts that automation implies on the industry revenues. Eventually, Citymobil type vehicles may be considered as yet another mode of public transport completing the PT offering.

About the needed *policy changes*, currently the public authorities face the urbanization and urban sprawl challenges translated in unacceptable levels of traffic congestion and increased need for liveable cities. As automation is brought to the market, authorities see many opportunities but also real threats if the technology is misused or used for individual purposes. It looks like the best solutions are not necessarily compatible with the individual needs. In this context, as politicians are under pressure to find solutions to make their cities more liveable, they try to understand how automation can contribute positively to the new needs to promote mobility as a whole i.e. in an integrated way. So, their key focus is to turn the automation progress at the service of the cities, i.e. of the community, rather than at the benefit of the individuals. This trend is in-line with the current city developments and urban design in many regions of Europe, which claim for:

- Keeping out the private cars from larger and larger portions of the city centre
- Favouring the soft modes and public transports
- Creating liveable areas surrounded by parking areas and crossed by public transport routes
- Reserving limited access to facility services (waste collection, delivery services, etc.).

The main issue in this top-down approach is the political willingness to create a livable environment at the cost of reducing the access to the users living outside the city. So one needs to work on the awareness of the benefits of livable cities and make sure that the demand comes from the citizens themselves rather than only through regulations. Then to make sure the outsiders can still access the city but in a controlled and livable way with more public transport and less cars. Fully automated transport has a clear role to play to convince the populations that manually driven vehicles do not belong to city centers.

4.3 Freight Operators

Freight discussion is difficult to fit in the scenarios discussed which are centered on passengers. In general, freight transport in cities is expected to be automated later than passenger transport because there is the need of a man to unload the parcels (the problem of last-meter transport). In Europe—this is no longer true in the US—there is also the need of a human to get the delivery document signed. Furthermore, the great difference in economic terms between passenger and freight transport is the little margins to which freight operators are working, which discourage to follow innovation for innovation sake; only when innovation will prove to be productive freight operator will invest in it.

The business model for the urban freight distribution is key. It is necessary to think to the adapted distribution chain already using Urban Distribution Centers. A push can come from sharing distribution vehicles to monetize assets (UBER_Freight Warehouse_B&B). It is necessary to make a distinction between delivering large quantities to large warehouses *versus* to small parcels to shops or private receivers. For the latter, it is possible to consider re-using the same (CityMobil2 like) vehicles for small parcels distribution (courier like). This would maximize the use of the vehicle and increase the investment return. A good example is provided by the city of La Rochelle, where there is already a mixed use of busses to deliver goods. This is a good way to develop a new integrated business model for transport. Besides the city logistic focus within the cities, urban distribution automation is seen as very positive also for long distance journeys. A new business model should therefore integrate long distance with urban distribution. The key enablers for long distance transport are:

- Removing the driver will extend operating hours and reducing operating costs (economy)
- Platooning will give advantages on fuel consumption (environment)
- Improved security.

About the needed *policy changes and key enablers*, there is a need to develop (or revise) the legislation on the driving/resting depending on the "auto-pilot" to monetize the positive effects of not having the driver engaged in driving tasks. The road code defines the following distance based on the human driving but platooning should enable shorter gaps, which would need to become legal. There would be the need to adapt the motorway maintenance program to the more precise driving, which will consume asphalt in precise location leaving it almost untouched elsewhere. A good enabler could be priority for freight delivery at traffic lights, which would save the pollution and road wear due to decelerating and accelerating a heavy vehicle. Finally, the tagging of the parcels is one of the key enablers as it would allow identification and automated retrieval of the package in the loading unit.

4.4 Private Operators

The session was attended by the few representatives of stakeholders' categories other than public authorities, automotive industry and freight transport operators. The discussion focused on the preferred scenario and, thanks to the presence of one stakeholder from the insurance industry, on issues of laws and liability.

As for the preferred scenario, both caricatural scenarios discussed in the workshop were considered as extreme cases, valuable to stimulate the discussion, not as realistic options. It is much more realistic a *hybrid scenario*, with a combination of privately owned and shared autonomous vehicles and still a portion of human driven cars even in the distant future. The main driver for the realization of the hybrid scenario will remain the private sector (companies and households), not the public, as only the former is considered to have in the future the assets to make the necessary investments. However, in the hybrid scenario there will be an important deployment and diffusion of *peer-to-peer sharing of autonomous vehicles*, also to help reducing the costs of the vehicles themselves, which are expected to be higher than the current prices of traditional cars (because the self-driving cars include more sophisticated equipment). The carmakers themselves will offer new multi-ownership options, to allow for example customers to buy self-driving vehicles only in the season they need to use them (it could be for summer holidays, for example, while in the working period the same customers could find convenient to commute with other modes). Peer-to-peer sharing will allow to abate directly the cost of car use, for example in case of ride sharing on commuting or long distance trips sharing the gasoline and toll costs with passengers, or by renting the car for a revenue in periods in which the owner does not need it.

As for the laws and liability issue, a major barrier to the full implementation of autonomous vehicles is legislation and governance. While Google has been testing its driverless technology in a fleet of cars for the past 3 years in the US and lobbying for new legislation, current European laws state that a person must be in control of a vehicle at all times. Autonomous vehicles also raise the question of liability. If these vehicles are safer and leave little room for error, other than potential mechanical or software glitches, who is responsible in the event of an accident? Is the technology company, the carmaker, or the occupant? By the same token, how alert will occupants need to be? This question is one many carmakers and technology providers are currently exploring. Many are looking for ways to keep the driver engaged since there will be some instances where the driver will need to be alert, or take control, particularly in the transition stages of autonomous technologies. There is also the risk that drivers' skills will reduce significantly with the use of more and more autonomous functions.

Finally, as it concerns the insurance issue, it is clear that with cars currently driven by humans there is a high risk of an accident due to the probability of human error. In the absence of human error, new forms of insurance will need to be devised. This necessity could be perceived as a barrier for insurance companies.

5 Conclusion

According to the majority of respondents urban transport automation will not cause in the key variables a radical change (within the range 10–30 %) in one direction or the other. This is because—in the opinion of many—autonomous vehicles, are only one of many factors that will affect transport demands and costs in the next few decades, and not necessarily the most important.

All *economic impacts* are expected to be positive in the private automated scenario, in particular in the urban sprawl context but also, with slight differences of intensity, in the other urban contexts (city network, small compact city, rural area).

The *social impacts* are expected to be positive in the collective automated scenario, for all urban contexts, while in the private automated scenario the impact on safety is assumed moderately negative, as the reduction of self-driving vehicle accident risk would be mitigated by a significant increase in the total mileage.

The *environmental impacts* are expected to worsen in the private automated scenario, due to the increased mileage not compensated by better vehicle and driving performances. In the shared self-driving transport scenario, the environmental impacts are always expected to improve—with the exception of infrastructure modification.

The *transport impacts* for the private automated scenario are expected to be respectively neutral (road capacity) and highly positive (comfort) in all the urban contexts.

The caricature scenarios discussed in the workshop were considered as extreme cases, valuable to stimulate the discussion, not as realistic options. It is much more realistic a hybrid scenario, with a combination of privately owned and shared autonomous vehicles and still a portion of human driven cars even in the distant future. The main driver for the realization of the hybrid scenario will remain the private sector, considered the strongest actor to make the necessary investments.

There is a need of *policy changes*.

It is clear indeed that autonomous vehicles are poised to be the next disruptive technology to travel. The challenge now for the world's city planners and managers is to understand how quickly autonomous vehicles will disrupt current patterns of passenger mobility, and if and how they may help public authorities to face the urbanization and urban sprawl challenges currently causing unacceptable levels of traffic congestion and an increased quest for more livable cities.

References

1. The study is our own original elaboration and it is part of the CityMobil2 project co-funded by the European Commission under the Framework Program 7
2. The complete report is available on the project website: http://www.citymobil2.eu/en/Downloads/Public-deliverables/

Synergies of Connectivity, Automation and Electrification of Road Vehicles

Gereon Meyer

Abstract The combination of connectivity and automation with the electrification of road vehicles offers a multitude of synergies in both performance of the technical systems and added values for users and businesses. These synergies become manifest in e.g. a higher energy efficiency and a more convenient operation. Furthermore they may define new products and services in the automotive domain. Therefore, they are an interesting subject of innovation analysis. This paper summarizes the activities of an international working group dealing with connectivity, automation, and electrification in road vehicles which was formed under the umbrella of the International Energy Agency (IEA). These activitites include the analysis of potential synergies, an information exchange about relevant research and development activities, and discussions on future trends in innovation, business development and deployment.

Keywords Electric vehicles · Car sharing · Automated and connected vehicles · Synergies

1 Introduction

According to roadmaps of the automotive industry, the implementation of high-degree automated driving (i.e. SAE levels 3 and above) can be expected for the motorway around the years 2020–2025, and in urban environments around 2025–2030 [1]. This coincides with the phase of broad market introduction of electric vehicles: A significant market share of such vehicles has been predicted for the 2020–2025 timeframe [2]. Interdependencies between the development and innovation processes in both technology fields are likely due to (a) similarities in the electronics and data architecture of control, (b) complementarities related to energy consumption, and (c) commonalities in the systematic character of the operating environment.

G. Meyer (✉)
Department Future Technologies and Europe, VDI/VDE Innovation + Technik GmbH, Steinplatz 1, 10623 Berlin, Germany
e-mail: gereon.meyer@vdivde-it.de

G. Meyer and S. Beiker (eds.), *Road Vehicle Automation 3*,
Lecture Notes in Mobility, DOI 10.1007/978-3-319-40503-2_14

Concept cars that are connected and automated on the one hand and electric on the other, are currently or will soon be under investigation in a multitude of field tests, e.g. EN-V 2.0 by General Motors is tested in Tianjin Eco-City (P.R. China), the Lutz pods will be available for trials in Milton Keynes (U.K.), and the use of driverless cars is studied by Google. Other examples include an electric delivery van that slowly follows the driver while he is walking from door to door as recently presented by Volkswagen, an automated valet parking and wireless charging service as proposed by Renault and other vehicle manufacturers, and electrified and driverless mobility robots as demonstrated by Hitachi.

In view of these developments, the Implementing Agreement (now: Technology Collaboration Programme) Hybrid and Electric Vehicles of the International Energy Agency recently established a new working group, Task 29 "Electrified, connected and automated vehicles". It analyzes the potential synergies in road vehicles, shares information about relevant research and development activities, and exchanges ideas on future trends in innovation, business development and deployment [3]. Some preliminary findings of the working group are summarized in the following paragraphs.

2 Synergies in Technology

The interdependencies of the technical functionalities of connected and automated electrified vehicles can be systematically analyzed by relating the enabling components and systems of both fields to each other. This is shown exemplarily in Fig. 1, where the existence of such interdependencies is indicated for the technology fields covered in the industrial R&I roadmaps that have been mentioned in the introductory paragraph.

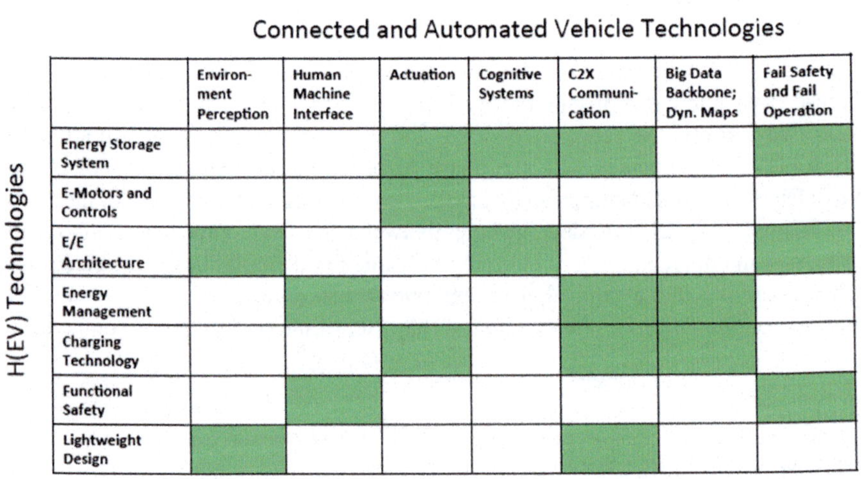

Fig. 1 Interdependencies between the key technologies of electric vehicles and connected and automated vehicles

In the following, a few of those interdependencies are described:

- **Fail Safety—E/E Architecture**: Electrified and automated vehicles are faced with comparable requirements for fail-operational design of the electric and electronic architecture—a great opportunity for exploiting synergies when designing the automobile of the future [4].
- **Actuation—Charging Technology**: Highly automated electric vehicles in combination with inductive charging can simultaneously find a parking spot with a charging-coil, position the vehicle coil accurately, and start the charging process automatically [5]. Future applications of this technology may even be dynamic, i.e. en-route inductive charging.
- **C2X Communication—Lightweight Design**: In the absence of mixed traffic and vulnerable road users, fully connected, SAE level 4/5 automated vehicles would be accident-free by definition. Hence, no passive safety measures would be needed to protect the occupants, and as a result the vehicles could be super-light. For an electric vehicle, this would mean a tremendous gain of range [6].

3 Complementarities in Energy Consumption

Connected and automated vehicles are able to choose routes and driving styles that minimize the energy consumption and ensure the best usage of the battery capacity in a hybrid or purely electric power train for a given road profile. Hence, an increased and predictable range of the electrified vehicle results. At the system level, automation in combination with cooperative driving ensures that traffic flows are optimized both in the city, the primary area of electric vehicle usage, and on the highway where it may greatly increase the usefulness of electric vehicles for longer distances.

However, due to the greater degree of convenience, automated vehicles may be driven more often and for longer distances. This may cause rebound effects to energy efficiency gains and to the synergies with electrification [7].

4 Role of Business Models

Despite the numerous opportunities for the exploitation of synergies due to the coincidence of connectivity and automation on the one hand and electrification on the other, it remains questionable whether these synergies will be used and are able to accelerate the adoption of those new road vehicle technologies. Not just higher purchase cost due to expensive sensor and energy storage systems may hinder an early adoption by private customers, but also the necessary investments in telecommunication infrastructure installations and energy storage systems, missing legal frameworks and hesitance of users may create hurdles. In the end, a connected, automated, and electrified road vehicle system might even be blocked than promoted.

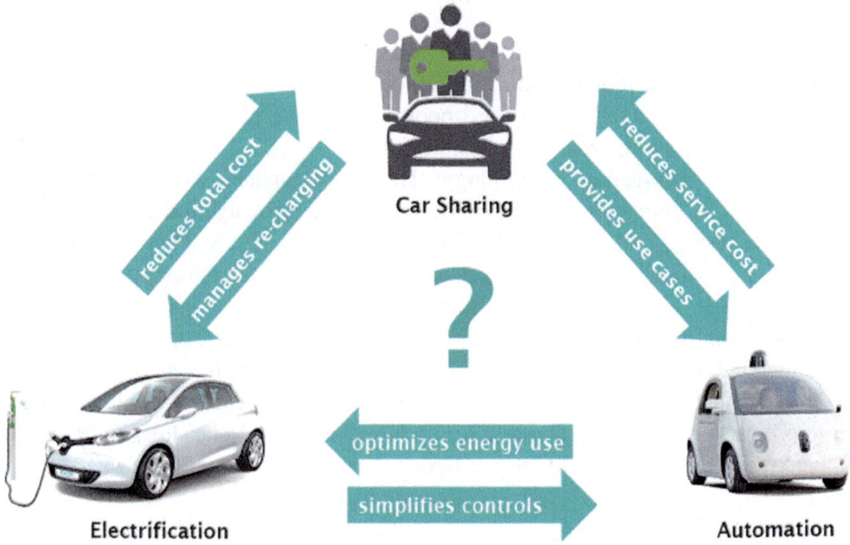

Fig. 2 Car sharing may help to unlock the potential synergies of electrification and connectivity/automation of road vehicles; the optimal vehicle design for this a subject of discussion

These issues may be solved through embedding the vehicles into a car sharing network because (a) the business model of car sharing acknowledges the fact that both electrified and connected/automated are cheaper in total cost of ownership than conventional vehicles, and (b) car sharing provides systemic functionalities like managed charging and on-demand availability that counteract limitations of individual ownership—yet another set of synergies (see Fig. 2).

5 Conclusions

In line with the findings of the IEA Implementing Agreement (now: Technology Collaboration Programme) Hybrid and Electric Vehicles' Task 29 "Connected and Automated Electrified Vehicles", the need for further research and innovation on synergies between the two fields has been promoted by the eNOVA Strategy Board Automobile Future in Germany, recently [8]. And, similar discussions are taking place in the framework of a task force "Electrification of Road Transport" at the European level currently.

From the preliminary findings, it can be stated that car sharing may play an important role in exploiting the synergies identified and in avoiding detrimental effects, e.g. in energy consumption. How the vehicle design can be optimized to fit all three domains, electrification, connectivity and automation, as well as car sharing, remains open, though.

References

1. European Technology Platform on Smart Systems Integration (EPoSS) (2015) European roadmap smart systems for automated driving. http://www.smart-systems-integration.org/
2. European Road Transport Research Advisory Council (ERTRAC), European Technology Platform on Smart Systems Integration (EPoSS), and SmartGrids, European Roadmap Electrification of Road Transport (2012). http://www.ertrac.org/
3. International Energy Agency Implementing Agreement Hybrid and Electric Vehicles, The Electric Drive Commutes, Annual Report, 2016 [online]. http://www.ieahev.org
4. Kuhn K-J (2014) Autonomous driving and electric mobility. Presentation at the EPoSS annual forum 2014
5. Automated Valet Parking and Charging for e-Mobility (V-Charge), Project Description (2015). http://www.v-charge.eu/
6. Rocky Mountain Institute (2011) Reinventing fire transportation sector methodology, Report, 2011
7. Brown A et al (2014) An analysis of possible energy impacts of automated vehicles. In: Meyer G, Beiker S (eds) Road vehicle automation. Springer International Publishing
8. Reuss H-C, Meyer G, Meurer M (2015) Roadmap 2030: Synergien von Elektromobilität und automatisiertem Fahren, ATZ Elektronik, 7

Part IV
Vehicle Systems and Technologies Development

Connected Truck Automation

Joshua P. Switkes and Steve Boyd

Abstract Commercial trucking is an industry ripe for connected and automated vehicles. The operations of the trucks combined with the highly analytical nature of the customers makes for the possibility of very rapid adoption. By combining partial automation with vehicle-to-vehicle and vehicle-to-cloud communication, these fleets can see massive fuels savings and safety improvements in the near term.

Keywords Platooning · Commercial trucking · Freight

1 Commercial Trucking: An Ideal Application of Connected and Automated Vehicles

Commercial trucking in the US and internationally is a huge industry that moves the vast majority of freight [1]. The vehicles are very flexible in their architecture because of the demands of the fleet customers, and these fleet vehicles are highly concentrated on major arteries (See Fig. 1). These factors together make trucking ideal for automation and connected vehicles.

1.1 Trucking

The trucking industry in the US alone is over $650 Billion [2]. Each truck in a long haul fleet may spend between 80 and $100 k per year on diesel fuel, with a large

J.P. Switkes (✉) · S. Boyd
Peloton Technology, 1060 La Avenida St, Mountain View, CA, USA
e-mail: Josh@peloton-tech.com

S. Boyd
e-mail: Steve@peloton-tech.com

G. Meyer and S. Beiker (eds.), *Road Vehicle Automation 3*,
Lecture Notes in Mobility, DOI 10.1007/978-3-319-40503-2_15

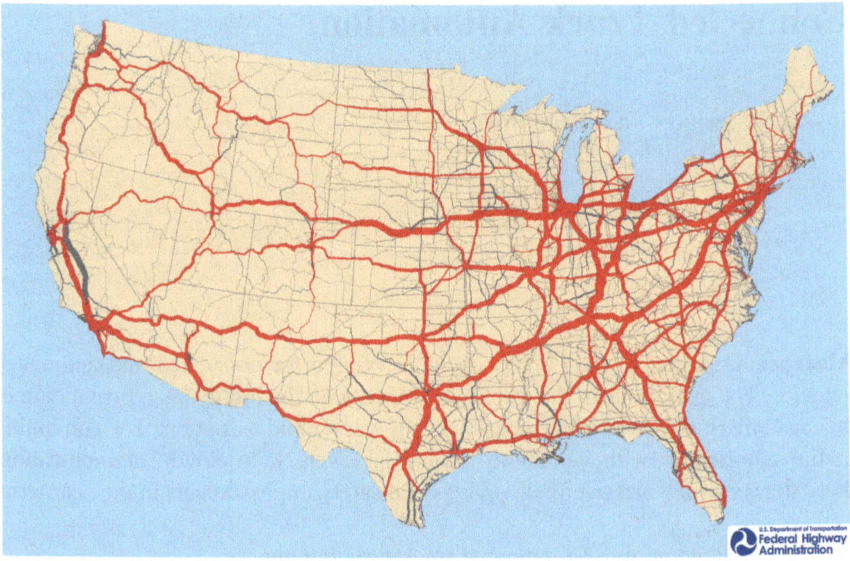

Fig. 1 Freight trucking concentration on major arteries [4]

fleet spending hundreds of millions of dollars a year or more on fuel [2].[1] Fuel is typically close to 40 % of the operating expense of a fleet, and the number one expense (American Transportation Research Institute 2015). These fleets, as many large established industries do, operate on very thin margins, typically 2–3 % net margins [3]. This means that savings on fuel can make a massive difference to their bottom line.

1.2 Flexible Vehicles

Heavy trucks are built upon a much more open and modular architecture compared to cars. Whereas a typical car will have a 3–5 year design cycle for each significant change, a truck platform will typically be around 10 years. Modifications requested by a fleet will be brought in in between these cycles. Fleets demand semi-custom trucks, with a typical order including not only the engine, the transmission, the brake supplier, but even features such as the wheelbase of the truck. To accommodate both the rapid design cycle, and this "mass-customization", the trucking industry has adopted features such as a standard for messaging: J1939. These allow the customization requested by fleets, and include messages needed for truck

[1]Calculated based on data from [2]. Average fuel cost per mile for truck-tractor was $0.64 in 2013; long haul tractors can drive 125–150 k miles per year and large fleets typically have over 10,000 tractors.

systems to interact, including engine operating parameters, chassis parameters such as wheel speed and steering angle, as well as supporting commands for engine torque and braking.

All together these messaging standards and the mass customization mean that new features can be introduced much more easily, and rapidly, on trucks compared to passenger vehicles.

1.3 Operations Well Suited to Platooning

Fleets are profit-driven operations, and as such the most successful are also highly organized. This has created several characteristics that are well suited to platooning and automation.

First, fleets do an excellent job of concentrating their operations on the roads that are most efficient for them. These are generally interstates and major US Highways. Figure 1 shows a map from the Federal Highway Administration, where the thickness of the line indicates the number of freight trucks per day.

This shows that trucks are very highly concentrated on this small number of routes and relatively small number of miles of highway. This is excellent for automation and platooning because it means that the roads that have to be considered when designing such a system can be constrained to this small number. It also means that trucks with cooperative/connected equipment will be near each other on the highway much more often than one might think when considering the millions of miles of road total in the US.

Many fleets also tend to have multiple trucks running together at the same time one each route. Many fleets have a high concentration of trucks on dedicated or line haul routes. For some fleets this is due to a planned hub-to-hub operation, while for others it is due to hub-and-spoke operation such as from a distribution center to stores. Other fleets have operations where the truck dispatch timing and routes are variable on a daily basis, due for example to customer requests. Some of these fleets still have many trucks running together due to a high regional concentration.

2 The Power of Connected Trucks

For the trucking industry there is great potential for connecting trucks to each other and to the cloud. Connecting them together overcomes some of the limitations of trucks, and connecting them to the cloud allows these systems to be constrained to certain operating conditions, which provides safety improvements and allows much more rapid deployment of the systems (Fig. 2).

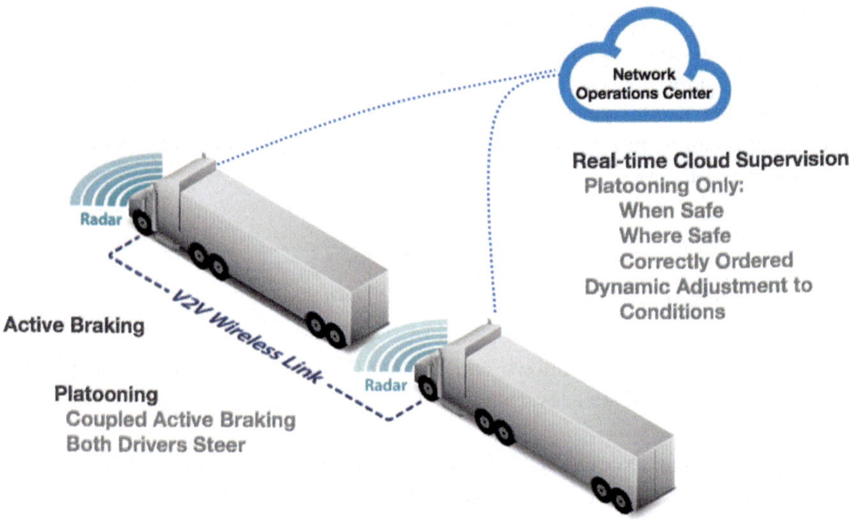

Fig. 2 Connecting trucks

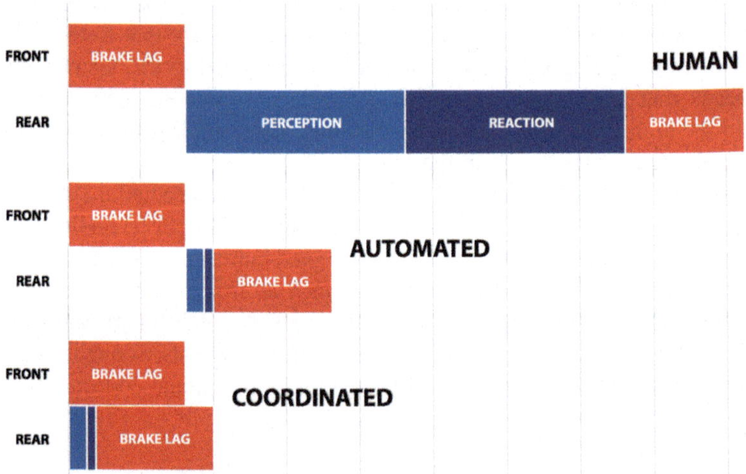

Fig. 3 Connected braking compared to manual and automated braking

2.1 Connected Braking

Connecting trucks to each other, through vehicle to vehicle communications, allows a direct method to overcome some of the limitations of heavy trucks. The biggest way is through connected braking.

Figure 3 shows three scenarios for a braking event between two trucks: Manual braking, automated braking, and connected braking. In each case the various time delays are represented by a bar graph, where each time delay corresponds to a distance at a given speed.

In the manual, or "Human" braking, a front truck applies the brakes, and after a brake lag the front truck starts to slow down. The driver of the rear truck can then perceive this slowing, and can then react to it. Finally the rear truck starts to slow after its brake lag.

This long, serial, process is the primary reason why truck drivers are instructed to follow at very long following distances: They are fundamentally limited in how well they can react the truck in front of them, due to both the characteristics of the truck brakes and their own human limitations.

By adding automation, such as a radar or lidar sensor, the human perception and reaction time/distance can be dramatically shortened, because the sensor is able to detect the slowing of the front truck before a human could perceive it, and because a computer can react much more quickly than a human can.

However this is still a serial process, where the perception and reaction can only occur once the brake lag of the front truck has elapsed. Only by coordinated/connected braking can this be eliminated. In this case the signal of braking is sent directly from the front truck to the rear truck upon *initiation* of braking. So even before the front truck has started to slow, the brakes can be applied in the rear truck.

This means that connected braking can provide a more immediate reaction than even the best automated system, because it overcomes one of the fundamental limitations of truck brake systems.

2.2 Cloud Connection

The second critical type of connection for truck automation/platooning is to connect the trucks to central cloud service. This is shown in Fig. 4.

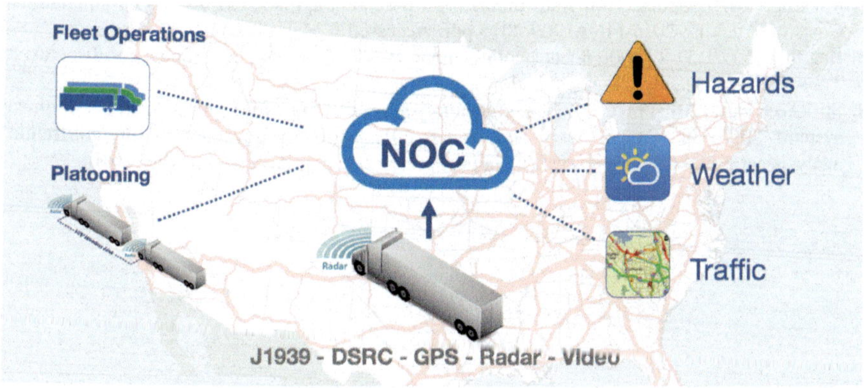

Fig. 4 Cloud connectivity to enhance platooning

The basic concept is to provide intelligence to the vehicle that would either be very difficult to determine from local information, or simply would be impossible to determine from local information.

For platooning and automation this means that we can restrict operations to only where it is safe or effective, when it is safe or effective, and how it is safe or effective. So we can restrict it to major interstates and US highways for example. We can adjust operating parameters such as following distance in platooning, or other parameters, in real-time. We can adjust these parameters based on information that trucks cannot possibly have from local sensors, such as upcoming traffic or weather information.

These types of operating restrictions mean that the development of these systems can be dramatically simplified. Rather than considering every possible type of road as one on which the system might be operated, it can be designed for only *known* roads. This makes for a much simpler validation process, and thus the systems can be deployed sooner than they would otherwise.

3 Conclusions

The trucking industry is ideally suited to platooning and automation due to the characteristics of their operation. Connecting trucks can also overcome limitations of the trucks themselves, and lead to a simpler and faster validation process.

References

1. American Trucking Association (2015) Trucking Revenues Top $700 Billion for the First Time According to New Report. http://www.trucking.org/article.aspx?uid=70210058-bb81-44df-a565-492f899fc139. Accessed 3 May 2016
2. American Transportation Research Institute (2015) An analysis of the operational costs of trucking: 2015 Update. http://atri-online.org/wp-content/uploads/2015/09/ATRI-Operational-Costs-of-Trucking-2015-FINAL-09-2015.pdf. Accessed 3 May 2016
3. Bearth DP (2015) TT100: fleets hauling more freight than ever but robust profits elusive. Transport Topics, July 22
4. Strocko E et al (2013) Figure 3-12 average daily long-haul truck traffic on the national highway system: 2011. In: Freight facts and figures 2013. http://ops.fhwa.dot.gov/Freight/freight_analysis/nat_freight_stats/docs/13factsfigures/figure3_12.htm. Accessed 3 May 2016

Validation and Verification of Automated Road Vehicles

Venkatesh Agaram, Frank Barickman, Felix Fahrenkrog, Edward Griffor, Ibro Muharemovic, Huei Peng, Jeremy Salinger, Steven Shladover and William Shogren

Abstract Ubiquitous, commercial deployment of automated road vehicles is desirable in order to realize their potential benefits such as crash avoidance, congestion mitigation, reduced environment impact, reduced driver stress, and increased driver productivity. A rigorous application of systems engineering, which includes validation and verification as crucial elements of assurance, is needed for the design and development of automated road vehicles. We discuss, without

V. Agaram (✉)
PTC Inc., 3310 West Big Beaver Road, Suite 100, Troy, MI, USA
e-mail: v.agaram@cimdata.com; vagaram@ptc.com

F. Barickman
National Highway Traffic Safety Administration, Vehicle Research and Test Center, 10820 SR 347, East Liberty, OH, USA
e-mail: frank.barickman@dot.gov

F. Fahrenkrog
Institut für Kraftfahrzeuge, RWTH Aachen University, Steinbachstraße 7, Aachen, Germany
e-mail: fahrenkrog@ika.rwth-aachen.de

E. Griffor
U.S. Department of Commerce, National Institute of Standards and Technology, 100 Bureau Drive, Gaithersburg, MD 20899-1070, USA
e-mail: Edward.griffor@nist.gov

I. Muharemovic
Continental Corporation, One Continental Drive, Auburn Hills, MI, USA
e-mail: ibro.muharemovic@continental-corporation.com

H. Peng
Michigan Mobility Transformation Center, Lay Auto Lab, 1231 Beal Ave, G036, Ann Arbor, MI, USA
e-mail: hpeng@umich.edu

J. Salinger
General Motors, 30500 Mound Road, Mail Code 480-106-RE2, Warren, MI, USA
e-mail: Jeremy.salinger@gm.com

G. Meyer and S. Beiker (eds.), *Road Vehicle Automation 3*,
Lecture Notes in Mobility, DOI 10.1007/978-3-319-40503-2_16

implying any form of joint recommendation, several areas of relevance to a common understanding of validation and verification of automated vehicles, namely customer expectations for vehicle response, industry standards for terms and definitions, industry standards for how measurement should be done, deeper knowledge of driving behavior today to serve as a reference, and standardized processes that encompass minimum performance requirements.

Keywords Validation and verification · Automated road vehicles · Commercial deployment · Systems engineering · Customer expectations · Industry standards · Terms and definitions · Driving behavior · Measurement standards

1 Introduction

The growth of sensors, actuators, and computational power in automobiles is facilitating the development of automated road vehicles but several crucial elements such as the customers' expectations, the road and traffic scenarios to be negotiated by the vehicles, the validation of customers' expectations, the verification of the functional and non-functional requirements, as well as the certification of the vehicles as roadworthy over their life, are still evolving. The validation and verification of automated road vehicles is not currently governed by standards or regulation although the National Highway Traffic Safety Administration (NHTSA) is actively working on the development of test protocols for lower levels of automation.

Systems engineering of road vehicles that fully or partially depend on the system performance for monitoring and assessing the hazards in the driving environment, for motion control such as steering, acceleration and deceleration, and for fallback dynamic driving tasks, is complex, from the technology integration perspective as well as from the perspective of dependable operation in complex road and traffic conditions. The uncertain road and traffic conditions make it very difficult to accurately identify the complex design envelopes of the automated vehicles. This in turn necessitates incorporation and integration of a large number of disparate technologies whose robustness in complex road scenarios is difficult to ensure over the life of the vehicles. Neither regulation nor comprehensive standards are available for such complex systems but a rigorous system engineering discipline must

S. Shladover
California PATH Program, University of California, 1357 South 46th Street, Building 452, Richmond, CA, USA
e-mail: sess@berkeley.edu

W. Shogren
Harman International, 39001 West 12 Mile Road, Farmington Hills, MI, USA
e-mail: William.shogren@harman.com

nevertheless be exercised to design for and deliver high performance and dependability.

Adherence to systems engineering discipline relies heavily on requirements validation, as well as on verification of functional specifications derived from those requirements. The identification of requirements however needs a good understanding of the full range of road and traffic scenarios in which the automated vehicles would operate. Some considerations would be highway driving in presence of conventional vehicles and vehicles with varying degrees of automation, versus driving in ambiguous conditions that exist on local roads with pedestrians, bicyclists, and often unclear signs and signals. The scenarios could be so diverse and complicated that lengthy learning programs would have to be established in order to conceive, tune, and refine practicable control algorithms for a commercially viable vehicle.

There is much unknown about customers' expectations, the set of road and traffic scenarios which are of relevance to automated vehicles, the tests that can represent the relevant scenarios, and the range of performance acceptable to different customer segments. It is not clear at which level of testing and tuning a prototype vehicle will be considered dependable or trustworthy enough for commercialization. In addition, it is unclear whether the automated vehicles will be accepted by the public as safe to use in daily life. Currently, individual carmakers, suppliers, and other companies are developing automated vehicles based on their understanding of what might be needed by tomorrow's customers. In the absence of performance standards, companies pursue their own methods of collecting data pertaining to the scenarios that they believe the automated vehicles would experience.

The authors of this chapter, who represent an automotive OEM, two automotive suppliers, a standards organization, three research organizations, and a regulatory body, have presented their views about validation and verification of automated road vehicles in an effort to achieve a common understanding and terminology that could facilitate communication about the challenges across the automotive and information technology industries, in order to make it easier to design, develop, and deploy automated vehicles.

2 The Validation and Verification Challenge

Historically, validation and verification has been done for vehicles that are much less complex than the automated road vehicles being developed today. Automated road vehicles need to be evaluated against human driving abilities, which is a much more difficult proposition. The main reason for this difficulty is that we don't fully understand how humans, in negotiating complex road and traffic scenarios, interact with conventional non-automated vehicles, and we cannot anticipate how humans will adapt, in the future, to different levels of automated vehicles. The question is "How can we achieve confidence that automated road vehicles will provide the desired value in both safety and travel convenience/utility"? Additionally, there are many “flavors” of automated driving systems, which include many unique aspects, each requiring extensive testing based on simulation, test track, and/or on-road evaluations. Moreover the simulations will have to include the entire vehicle system, representative operating scenarios and representative human response. Essentially, two types of guidelines have been created to help build confidence in active safety products, namely, those which focus on the processes used for development and testing, and those which focus on objective tests at all levels for each feature.

Due to the fact that automated vehicles will have capabilities which are a significant departure from today’s production vehicles, it is necessary to create social norms for the response of such vehicles, i.e., common expectations for how automated vehicles maneuver and interact with others in their vicinity, as well as, consistent terminology for roadway and environmental conditions, test conditions, and performance metrics. These norms might be established through cross-industry precompetitive cooperation in relevant technical work groups. They could then be used to guide performance testing for behavior, sensor interference, and security.

Different levels of automated vehicles find themselves at different experimental stages of development. Considering the evolutionary development of automated driving, the fundamental knowledge of sensing reliability, performance and safety is improving. However, as different active safety technologies are combined and automated driving features are developed, the learning curve increases exponentially. Consequently, redundancies and fallbacks are often architecturally developed based on that learning. Currently, a significant amount of time is invested in developing not only novel methods but also self-learning algorithms for testing the automated driving functions that are necessary to deliver safe, reliable, secure and robust self-driving vehicles. Consequently, as the level of automation increases, considerations of system failures need to include those functions that can be transferred to the driver (“fail silently”) and those that cannot be transferred to the driver (“fail operationally”).

Validation and verification of automated road vehicles is bounded by four main areas: (i) the human-machine interface which governs how the vehicle and the driver interact, (ii) the customer-satisfaction which covers how comfortable the driver feels, (iii) the cybersecurity of the invariably connected automated vehicles, and finally, (iv) the operating environment which is made up of an extremely large number of driving scenarios. Also, considering the paucity of experiential data, the robustness of automated vehicle designs of higher levels of automation can be improved by collecting the performance data from semi-automated vehicles after deployment, in order to identify “rare events” and hidden risks. This knowledge can potentially be shared across the industry on a precompetitive basis, to facilitate more capable and safe vehicles sooner.

Vehicle regulators like the NHTSA aim to implement testing protocols that ensure minimum performance in automated road vehicles. The NHTSA is surrogate for a vehicle owner/occupant and other road users. Its objective is to regulate testing to validate the minimum level of safety that the automated road vehicles must provide. In this regard, the safety principles proposed by the Crash Avoidance Metrics Partnership (CAMP) [1–3] could be a viable starting point for the development of validation and verification of automated road vehicles, beginning with objectively determining the level of automation of the vehicle.

The NHTSA has been developing protocols for testing lower level automation functions like park-assist and has been working towards developing a systematic way of applying that methodology to higher levels of automation. Safety test protocols need to cover the reliable functioning of sensors, the ease of understanding of the human interfaces, and demonstrate that the safety risk is acceptable. The main complexity today resides in different automated vehicle systems designed with different interfaces, control systems, and functioning paradigms which can be very confusing.

Validation and verification of automated road vehicles faces significant challenges today due to the variation in how different systems respond to similar situations, sometimes conflicting with the expectations of either their drivers or the drivers of other vehicles. Also, the communication with the drivers via chimes, cluster indications, etc., needs to be "implicit" and be able to address the dependencies between overlapping information [4]. Further, the automated functions cannot be tested for all operating scenarios, but must be expected to perform within safety guidelines in all scenarios.

Most of the automation approaches use localization (understanding of the vehicle state relative to its surroundings) achieved through onboard systems because the integrity of the off-board data may be lacking. Though localization is critical to situational response, it cannot be tested for all possible situations.

Finally, automation below level 3 will require a model of the driver as part of the "situation" and that also needs incorporation into testing considerations.

In terms of regulation of automated road vehicles, one of the challenges is striking a balance between public safety and encouragement of innovation when technical standards do not exist. At the level of the states in the USA, manual driving is regulated through registration and driver licensing. Should this model be extended to automated road vehicles, then that could form the basis for validation and verification, keeping in perspective that it needs to be achieved at an affordable price and in an unambiguous way. One could apply functional safety principles manifested in ISO 26262 as guidelines for developing dependable automated vehicles. However, that standard would have to be augmented with a pass/fail criterion to be useful in a regulatory framework, which in turn would need considerable research and data. Also, some have suggested that results of the validation tests should be reportable to interested members of general public to enhance public confidence in the safety of the systems, but this needs to be traded off against protection of developers' intellectual property. Finally, third party safety process review or safety design review can also be used to enhance public confidence but it adds undue expense and has significant IP issues.

The definition of validation [5] is—"The assurance that a product, service, or system meets the needs of the customer and other identified stakeholders. It often involves evaluating acceptance and suitability with external customers." The definition of verification [5] is—"The evaluation of whether or not a product, service, or system complies with a regulation, requirement, specification, or imposed condition. It is often an internal process." Verification is part of the technical assessment (checking requirements)—methodology for impact analysis of automated driving applications in the European research project AdaptIVe. Validation is part of safety and environmental impact assessment as well as the user-related, in-traffic and technical assessment.

The automated driving functions can be classified based on the time of operation into—(i) Event based—function that operates for a short period of time (typically vehicle stands still at the end, i.e., the automated driving ends), and (ii) Continuously operating—function that operates for a longer period of time (typically vehicle is still moving at the end of a maneuver, i.e., automated driving continues). Different evaluation approaches for event-based and continuously operating function can be foreseen. Independent of the chosen approach the main evaluation criteria for AdaptIVe are that the automated driving systems need to operate within the range of normal driving behavior (i.e., not disturb normal driving in mixed traffic) and should at least be as safe as non-automated driving.

The impact assessment of automated driving applications is constrained due to—(i) today's accident data not including collision of automated vehicles, (ii) accident reconstruction rendered more difficult to automated vehicle function, and (iii) the not-yet-understood complex interaction between automated and non-automated vehicles.

Safety impact assessment can be a three step approach based on (i) identification of relevant scenarios, focused on crashes and other (relevant) driving situations, (ii) investigation of relevant scenarios in detail similar to scenario reconstruction approach, and (iii) identification of new scenarios such as transition of control or minimum risk maneuver.

There are at least five different approaches of evaluation that can be applied to capturing the operating environment of automated road vehicles, namely, the test matrix approach, the naturalistic field operations approach, the Monte Carlo simulations approach, the worst-case evaluation approach, and the accelerated evaluation approach. The most likely predictor of system behavior is a full system simulation against the characterization of the operating environment provided by all of these. Though multiple vehicle subsystems have well developed simulations there are serious obstacles to integrating them into a single whole system simulation. Efforts to provide such a 'federated' simulation environment are in process at NIST and a number of academic institutions, including UC Berkeley and Vanderbilt University.

The test matrix approach is repeatable, easy to execute, and fast, which the FMVSS and NCAP will likely continue to use although it does not ensure learning, the selection of the matrix is somewhat arbitrary, and "scoring" does not relate to real-world safety benefit. The naturalistic field operational testing, which is being used by Google, is directly related to the real world but is a slow and expensive method with low exposure to safety critical scenarios. The Monte Carlo simulation approach, which is based on the data from the naturalistic field operation testing, is suitable for simulations and driving simulators but is not amenable to accelerating the scenarios. The worst case evaluation approach explores and focuses on weakness, which is an advantage but it needs vehicle models and control models, and is numerically challenging. Moreover, its relationship with the real-world scenarios is not clear. The accelerated evaluation approach takes the naturalistic driving data along with the disturbance model of the behavior of other vehicles, and skews the disturbance statistics to accentuate the portions of interest and skews back to understand the real-world safety benefits. This can accelerate the testing or simulations by 100–10,000 times and still compute real-world safety benefits although it needs a large quantity of driving data to begin, which can be an area of pre-competitive collaboration.

3 Paving the Way for Validation and Verification

Validation and verification of automated road vehicles will be governed by safety and security, as well as by performance, comfort and convenience. Given the fact that many of the functions at different levels of automation are in different experimental stages, and will need a long time before they can be commercialized, a collaborative effort between OEMs, suppliers, research organizations, standards organizations, and regulatory bodies could help remove some easily removable obstacles. Some relevant research questions related to the validation and verification topic are:

1. Is crash safety the dominant function for validation and verification? How does verification and validation of other functions e.g., passenger comfort, compare in terms of understanding and complexity?
2. What are the roles of standards and perhaps regulation in the area of validation and verification? Would they be beneficial or would they be an impediment to innovation?
3. Can we follow the example of other technology driven functionalities in order to develop a template for validating and verifying automated vehicles and eventually standards?
4. Is there an opportunity to identify automated vehicle subsystems for a standards exercise or precompetitive exchange without seriously impacting competitive advantage?
5. Would it be possible to share road study data in a precompetitive manner to improve verification and validation?
6. How can we improve communication about systems and components between OEMs and suppliers to improve the efficiency of verification and validation of automated vehicles? Example—common definitions, terminology, etc.
7. Would it be possible to consider standards for the operating environment for verification and validation? Example—weather, road conditions, traffic scenarios, etc.

8. What are appropriate test tools (field, test track, simulation, HIL) for validation and verification? Do we need new test tools or are the existing test tools sufficient?
9. Are different approaches needed for different types of automated driving functions (automated parking function vs. highway automated driving function or V2V functions vs. non cooperative functions)?

The opinions of the authors in the context of these questions are presented in the following.

Validation and verification can involve customer satisfaction, human-machine interface, cybersecurity, and challenging operating environments. Some have suggested that automated vehicles must achieve the overall status that is no worse than when humans drive cars. The change in traffic patterns due to introduction of automated road vehicles cannot be fully predicted.

Safety is the highest priority. Technology introduction should not make road traffic unsafe. Moreover, there is a fine balance between safety features and ease of use—the potential safety benefit is diminished if customers shut off a function due to annoyance or due to lack of understanding.

Systems engineering needs to cover safety engineering and security engineering, adding complexity. The level of system complexity governs the complexity of breaking down verification and validation into different tests. Systems working differently but performing the same overall function create additional validation and verification complexity. The potential for sensor algorithm errors creates a new level of complexity for validation and verification. Finally, validation and verification of automated vehicles is much more complex because its reference is the ability of what a human does or can do, and that in itself is not well understood. More knowledge about normative human behavior and expectations would improve validation.

Acceptance tests are expensive due to the large number of options. It is challenging to do validation and verification cost effectively. Mixed traffic conditions must be taken into account for validation and verification but for cost reasons vehicle manufacturers and their suppliers may need to restrict themselves to microscopic traffic simulation. Manufacturers may need to consider "worst case scenario" and "accelerated evaluation" in order to reduce the cost of validation and verification yet cover unusual scenarios that will occur in real world.

Common industry standards around terms and definitions help the development process. Examples include common terminology for driving situations, environments, road classes, and traffic situations, etc. Precompetitive collaboration between industry, government, and research organizations on social norms and expectations of road users, consistent terminology, and shared data quality assurance is desirable.

Common industry standards around how measurement should be done would help improve communication and understanding. The driver of an automated vehicle needs to know when and how he or she needs to take back control. Some have suggested this should be consistent from vehicle to vehicle.

If road and traffic scenario data is to be shared among OEMs, suppliers, and other organizations, then a common understanding of what that data should be and how to interpret it needs to be established. Such joint data could potentially help promote better understanding and acceptance, and inform decisions about policy and not just technical issues.

Data collection from vehicles after launch is an opportunity to identify rare events in order to expose hidden risks that don't occur in pre-release testing, and this could be an area of precompetitive data sharing without risking intellectual property.

Simulation and modelling is becoming more and more reliable but we still need physical testing to be confident, particularly when dealing with safety certification. Consequently, it will be used in the development domain but unlikely to enter the certification arena in the near future, given the complexity involved.

4 Conclusion

The authors have discussed validation and verification of automated vehicles from the perspectives of manufacturers, suppliers, researcher organizations, standards organization, and regulation, and find that the main areas for future exploration, in order to make it easier to design, develop, and deploy automated vehicles, should be driven by the following needs:

1. Development of common industry terms and definitions.
2. Creation of common customer expectation for automated vehicle response.
3. Development of common industry standards around how performance should be measured.
4. Better knowledge of driving behavior as it is today from the data already available.
5. Collection of post-launch data to help identify rare events.
6. Increased use of modeling and simulation to reduce verification costs.
7. Incorporation of driver models into verification and validation.

Acknowledgments The authors would like to acknowledge Mary Doyle of the Society of Automotive Engineers for capturing the details of the breakout session on Verification and Validation of On-Road Automated Vehicles held at the Automated Vehicle Symposium 2015. The first author would like to acknowledge Paul Perrone of Perrone Robotics for preparing the initial ground for the breakout session on Verification and Validation of On-Road Automated Vehicles held at the Automated Vehicle Symposium 2015.

References

1. Barrickman F (2014) USDOT-crash avoidance metrics partnership automation research project overview. SAE Government and Industry Meeting, Washington, D.C
2. Christensen A, Cunningham A, Engelman J, Green C, Kawashima C, Kiger S, Prokhorov D, Tellis L, Wendling B, Barickman F (2015) Key considerations in the development of driving automation systems, 24th enhanced safety vehicles conference. Gothenburg, Sweden
3. Tellis L (2015) Key considerations in the development of driving automation systems. Automated Vehicles Symposium, Ann Arbor, Michigan
4. Griffor ER, Nass C (2016) Implicit communication: design for human-machine interaction, handbook of system safety and security, Editor Griffor, E.R., Elsevier-North Holland Publishing, 2016 (to appear)
5. 1490 WG—IEEE Guide—Adoptions of the Project Management Institute (PMI) Standard—A Guide to the Project Management Body of Knowledge (PMBOK Guide), 4th edn, 2008

Trustworthy Foundation for CAVs in an Uncertain World: From Wireless Networking, Sensing, and Control to Software-Defined Infrastructure

Hongwei Zhang, Le Yi Wang, George Yin, Shengbo Eben Li, Keqiang Li, Jing Hua, Yeuhua Wang, Chuan Li and Hai Jin

Abstract Three basic enablers for connected and automated vehicles (CAVs) are wireless networking, sensing, and control. Tightly coupled with the physical process of wireless signal propagation, vehicle movement, and environment, however, CAV wireless networking, sensing, and control are subject to complex cyber-physical uncertainties. To address the challenges, we propose an integrated, cross-layer framework for taming cyber-physical uncertainties, within which we develop novel algorithms and methodologies for addressing the interdependencies between networking, sensing, control, and physical processes. To enable high-fidelity evaluation and thus the deployment and adoption of new CAV tech-

H. Zhang (✉) · J. Hua · Y. Wang · C. Li · H. Jin
Wayne State University, 5057 Woodward Ave., Detroit, MI 48202, USA
e-mail: hongwei@wayne.edu

J. Hua
e-mail: jinghua@wayne.edu

Y. Wang
e-mail: yuehua@wayne.edu

C. Li
e-mail: chuan@wayne.edu

H. Jin
e-mail: haijin@wayne.edu

L.Y. Wang
Wayne State University, 5050 Anthony Wayne Dr., Detroit, MI 48202, USA
e-mail: lywang@wayne.edu

G. Yin
Wayne State University, 656 West Kirby, Detroit, MI 48202, USA
e-mail: gyin@wayne.edu

S.E. Li
Tsinghua University, A643 LSK Building, Tsinghua Campus, Beijing 100084, China
e-mail: lishbo@tsinghua.edu.cn

K. Li
Tsinghua University, 201 ARI, Tsinghua Campus, Beijing 100084, China
e-mail: likq@tsinghua.edu.cn

G. Meyer and S. Beiker (eds.), *Road Vehicle Automation 3*,
Lecture Notes in Mobility, DOI 10.1007/978-3-319-40503-2_17

nologies, we develop a software-defined CAV infrastructure for conducting CAV experiments using vehicles in real-world traffic so that properties of V2X communication, vehicles, traffic, road, and environment are captured at high-fidelity.

Keywords CAV · Wireless networking · Sensing · Control · Software-defined infrastructure

1 Introduction

Transforming the traditional, single-vehicle-based safety and efficiency control, next-generation vehicles will cooperate with one another and with transportation infrastructures to ensure safety, maximize fuel economy, and minimize emission as well as congestion. Three basic enablers for this vision of connected and automated vehicles (CAVs) are wireless networking, sensing, and control: wireless networking enables real-time sensing and control within the ecosystem of vehicles, infrastructures, and environment; networked sensing enables comprehensive, beyond-line-of-sight sensing for optimal CAV control; networked control enables intelligent action upon the real-time information across the ecosystem of infrastructures and vehicles [1].

Tightly coupled with the physical process of wireless signal propagation, vehicle movement, and environment, however, CAV wireless networking, sensing, and control are subject to complex dynamics and uncertainties in both the physical domain and the cyber domain. In the *physical* domain, multipath signal propagation and power attenuation introduce anisotropy, asymmetry, and complex spatiotemporal dynamics in wireless communication; together with wireless interference, they introduce uncertainties in the reliability, timeliness, and throughput of wireless communication. Vehicle mobility introduces dynamics in vehicle spatial distribution, thus further increasing the uncertainty in wireless channel properties. Additionally, CAV sensing and control are subject to uncertainties in the physical environment such as bad weather, sharp turns, and location-specific vehicle traffic pattern. In the *cyber* domain, dynamics in wireless networking, sensing, and control interact with one another during their adaptation to physical dynamics. For instance, the real-time capacity (i.e., amount of data deliverable within a certain time duration) of wireless networking changes during its adaptation to physical-domain dynamics; accordingly, CAV sensing and control adapt their optimal strategies to the in situ network real-time capacity. Dynamic sensing and control strategies in turn generate dynamic network traffic pattern (e.g., load) and pose dynamic requirements on data delivery reliability, timeliness, and throughput. These complex dynamics and uncertainties, together with the real-time, safety-critical nature of CAV sensing and control, require us to rethink the theory and practice of wireless networking as well as vehicle sensing and control.

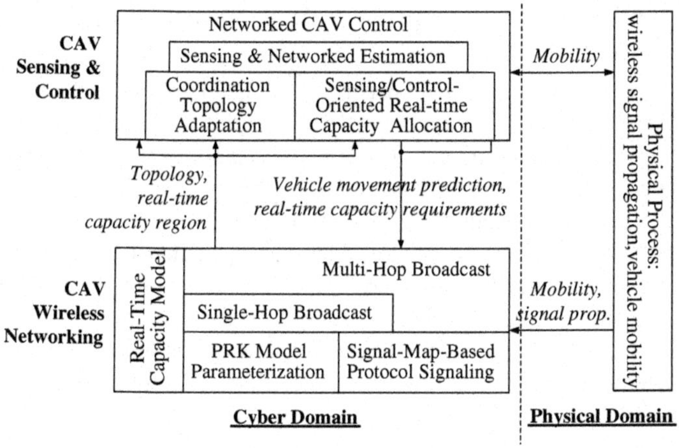

Fig. 1 Integrated, cross-layer framework for CAV wireless networking, sensing and control

To address the challenge, we propose an integrated, cross-layer framework for taming cyber-physical uncertainties in CAV wireless networking, sensing, and control as shown Fig. 1. In this framework, wireless networking, sensing, and control interact with one other to address cyber-physical uncertainties. Based on the real-time capacity region of wireless networking and the physical process of vehicle movement, traffic, and environment, CAV sensing and control select their optimal strategies and the corresponding requirements on the timeliness and throughput of wireless data delivery (e.g., for maximizing sensing accuracy and roadway utilization while ensuring safety). Based on the requirements from CAV sensing and control, wireless networking adapts to cyber-physical uncertainties to ensure the timeliness and throughput of V2X communication; for addressing the impact of vehicle mobility on wireless communication, wireless networking also leverages input from vehicle control on vehicle movement prediction.

As new CAV technologies are developed, they need to be evaluated before their real-world deployment and adoption. The impact of vehicle traffic, road, and environmental conditions (e.g., buildings) on CAV wireless communication and the safety-critical nature of vehicle operation require evaluating CAV technologies and applications in real-world settings. Thus it is desirable to conduct CAV experiments using vehicles in real-world traffic so that properties of V2X communication, vehicles, traffic, road, and environment are captured at high-fidelity. To realize this paradigm of symbiotic CAV experiments in real-world traffic, we develop a software-defined CAV infrastructure through software-defined virtualization (SDV) and multi-domain emulation (MDE) of CAVs.

The rest of this chapter is organized as follows. In Sects. 2, 3, and 4, we elaborate on our recent progress in CAV wireless networking, control, and software-defined infrastructure, and we make concluding remarks in Sect. 5.

2 Predictable CAV Wireless Control Networking

In CAV wireless networking, the cyber-physical uncertainties challenge the models and protocols for scheduling concurrent transmissions to control co-channel interference, which is a basic issue in wireless networking and affects the predictability in data delivery reliability, timeliness, and throughput [2, 3]. For agile adaptation to dynamics and uncertainties, distributed scheduling is desirable in vehicular wireless networking. Nonetheless, most existing literature are either based on the protocol interference model or the physical interference model, neither of which is a good foundation for distributed interference control in the presence of dynamics and uncertainties [2, 3]. To address this issue, we have proposed the *physical-ratio-K (PRK)* interference model that integrates the protocol model's locality with the physical model's high-fidelity [2].

In the PRK model, a node C' is regarded as not interfering and thus can transmit concurrently with the transmission from another node S to its receiver R if and only if

$$P(C', R) < \frac{P(S, R)}{K_{S,R,T_{S,R}}}$$

where $P(C', R)$ and $P(S, R)$ is the average strength of signals reaching R from C' and S respectively, $K_{S,R,T_{S,R}}$ is the minimum real number (i.e., can be non-integer) chosen such that, in the presence of interference from all concurrent transmitters, the probability for R to successfully receive packets from S is no less than the minimum link reliability $T_{S,R}$ required by applications (e.g., CAV control). As shown in Fig. 2, the PRK model defines, for each link (S, R), an exclusion region $\Lambda_{S,R,T_{S,R}}$ around the receiver R such that a node $C \in \Lambda_{S,R,T_{S,R}}$ if and only if $P(C', R) \geq \frac{P(S,R)}{K_{S,R,T_{S,R}}}$. Accordingly, every node $C \in \Lambda_{S,R,T_{S,R}}$ is regarded as interfering with and thus shall not transmit concurrently with the transmission from S to R.

For enabling predictable interference control in the presence of network and environmental uncertainties, the parameter $K_{S,R,T_{S,R}}$ of the PRK model adapts to the

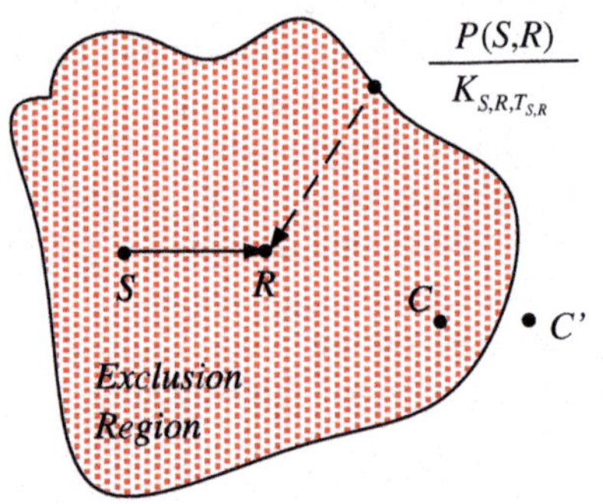

Fig. 2 Physical-ratio-K (PRK) interference model

specific network and environmental conditions to ensure the application-specific link reliability requirements. By ensuring the required link reliability and by using signal strength instead of geographic distance in model formulation, the PRK model captures the properties of wireless communication (e.g., cumulative interference and anisotropic signal propagation) and thus is of high-fidelity. For enabling distributed protocol design and implementation, the PRK model is also local [2, 3]: (1) The parameters of the PRK model are either locally measurable (i.e., for the signal strength and link reliability between close-by nodes) or locally controllable (i.e., for $K_{S,R,T_{S,R}}$ of each link (S, R)), thus PRK-based scheduling does not need to rely on parameters such as nodes' locations or channel path loss between far-away nodes which are often used in physical-model-based scheduling but are difficult to obtain precisely, especially in a distributed manner; (2) Only pairwise interference relations between close-by nodes need to be defined in the PRK model, thus PRK-based scheduling does not require explicit global coordination which is often used in physical-model-based scheduling. Through comprehensive analysis, simulation, and measurement, we have observed that, by ensuring the required link reliability, PRK-based scheduling also helps reduce data delivery delay by minimizing the need for packet retransmissions; we have also found that PRK-based scheduling can enable a channel spatial reuse very close to (e.g., >95 %) what is feasible in physical-model-based scheduling while ensuring application-required reliability [2]. Therefore, the PRK model serves as a good foundation for predictable interference control.

A basic task in PRK-based scheduling is to instantiate the PRK model, i.e., identifying the parameter $K_{S,R,T_{S,R}}$ for every link (S, R), according to in situ, potentially unpredictable network and environmental conditions. It is, however, difficult to characterize the relation between $K_{S,R,T_{S,R}}$ and the packet delivery reliability along (S, R) in closed-form, and the relation is complex and dependent on network and environmental conditions which may well be unpredictable at design time [2]. To address the challenge, we observe that the PRK model instantiation problem can be formulated as an online *regulation control* problem [4], where the "plant" is the link (S, R), the "reference input" is the required link reliability $T_{S,R}$, the "output" is the actual link reliability $Y_{S,R}$ from S to R, and the "control input" is the PRK model parameter $K_{S,R,T_{S,R}}$. Then we can leverage minimum-variance regulation control theory [4] to derive the controller for instantiating the PRK model parameter [3].

Given the instantiated PRK model parameter $K_{S,R,T_{S,R}}$ and using the *local signal map* that contains the average signal power attenuation between R and every node C close-by, R can identify the set of nodes whose transmissions can interfere with and thus cannot be concurrent with the transmission along link (S, R) [3]. With precise awareness of mutual interference relations with close-by nodes/links, nodes schedule data transmissions in a TDMA fashion using our distributed Optimal-Node-Activation-Multiple-Access (ONAMA) algorithm [5] to avoid concurrent transmissions along interfering links [3]; for convenience, we denote the resulting PRK-based scheduling protocol as PRKS [3].

Through extensive measurement study in the high-fidelity Indriya and NetEye wireless network testbeds, we observe the following [3]: (1) The distributed controllers for PRK model instantiation enable network-wide convergence to a state where the desired link reliabilities are ensured; (2) With local, distributed coordination alone, PRKS achieves a channel spatial reuse very close to what is enabled by the state-of-the-art centralized physical-model-based scheduler iOrder [6] while ensuring the required link reliability; (3) Unlike existing scheduling protocols where link reliability is unpredictable and the ratio of links whose reliability meets application requirements can be as low as 0 %, PRKS enables predictably high link reliability (e.g., 95 %) for all the links in different network and environmental conditions without a priori knowledge of these conditions; (4) By ensuring the required link reliability in scheduling, PRKS also enables a lower communication delay and a higher network throughput than existing scheduling protocols. Therefore, PRKS serves as an effective, field-deployable solution to predictable interference control, which has been an open problem for over 40 years [7], and PRKS enables predictable communication reliability which is a foundation for predictable CAV wireless networking and control in general.

3 CAV Control

We employ platoon control to discuss CAV control issues. The early studies of platoons, such as the PATH program in California in 1980s, targeted many fundamental topics, including goals, task division, control architectures, sensing and actuation, and control laws for headway control, etc. [8]. Since then, broader issues have been pursued, such as spacing policies, powertrain dynamics, and the impact of homogeneity and heterogeneity; with real world demonstrations, exemplified by GCDC in Netherlands, SARTRE in Europe, and Energy-ITS in Japan.

The earlier platoons employed radar-based sensing systems with highly limited information exchange topologies. The rapid deployment of V2X communications can accommodate various types of information topologies, e.g., two-predecessor following type and multiple-predecessor following type. New challenges naturally arise due to topology varieties, communication time-delay, packet loss, and quantization error. From a cyber system viewpoint, a vehicular platoon is a networked dynamical system with distributed controllers. Within this framework, vehicles in a platoon use their neighborhood information for controller design but must coordinate to achieve a global goal. As proposed by Li et al. [9, 10], such a perspective naturally decomposes a platoon system into four interrelated components as shown in Fig. 3: (1) Node dynamics (ND), which describes the behavior of each vehicle. The vehicle longitudinal dynamics are represented by nonlinear models of engine, drive line, brake system, aerodynamics drag, tire friction, rolling resistance, gravitational force, etc. They are often simplified to linear models in practice, e.g., single integrator model, second-order model (including double-integrator model), third-order model, and single-input-single-out model; (2) Information flow

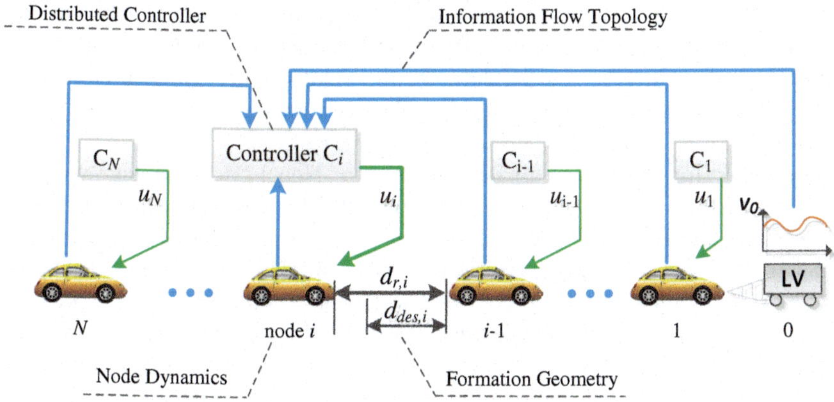

Fig. 3 Four major components of a platoon system: node dynamics, information flow topology, formation geometry, and distributed controller. $d_{r,i}$ is the actual relative distance,$d_{des,i}$ is the desired distance, u_i is the control signal for the i-th vehicle, and C_i denotes the controller in the i-th vehicle

topology (IFT), which defines how the nodes exchange information with each other. The IFT is usually represented and studied by using algebraic graph theory; (3) Formation geometry (FG), which dictates the desired inter-vehicle distances. There exist three major policies of FG: constant distance policy, constant time headway policy, and nonlinear distance policy. The objective of platoon control is to track the speed of the leading vehicle while maintaining a formation governed by the desired spacing policy between consecutive vehicles; (4) Distributed controller (DC), which implements the feedback control using only neighboring information; At present, most DCs are linear for rigorous performance analysis and hardware implementation. Since internal stability of the closed-loop system depends critically on IFTs, linear DC design is often case-specific.

Linear DCs suffer from difficulties in explicitly ensuring string stability or accommodating state or control constraints. Recently, $\mathcal{H}_\infty$ controller synthesis has been proposed to include the string stability requirement as a priori design specification. In addition, model predictive control (MPC) has been introduced to forecast system dynamics, explicitly handling actuator/state constraints by optimizing given objectives.

Within this four-component platoon cyber-physical system, platoon control aims to maximize highway utility while ensuring zero accident. To increase highway utility, it is desirable to reduce inter-vehicle distances and to accelerate fast towards a new formation after disturbances. These, however, will increase the risk of collision in the presence of vehicle traffic uncertainties. This tradeoff prompts a systematic design to *maximize benefits at a tolerable risk*. Interestingly, this platoon performance problem bears striking similarity to financial portfolio management problems aiming to maximize profit returns with controlled risk, which have been rigorously studied in the mean-variance (MV) framework. Recently, we have applied the MV method to study platoon control [11]. The MV method offers distinct advantages: (1) unlike heuristic methods such as neural networks and genetic algorithms, the MV method is simple but rigorous; (2) the MV method is computationally efficient; (3) the form of the solution (i.e., efficient frontier) is readily applicable to assessing risks in platoon formation, hence is practically appealing.

The MV framework originated from the Nobel-price-winning work of Markowitz [12]. In finance, it enables an investor to seek highest return (mean) under a pre-specified risk level (variance of the return). It has developed into a foundation of modern finance theory and been applied to other fields. Using the stochastic linear-quadratic (LQ) control framework, Zhou and Li [13] studied the MV in continuous-time systems, in which the control weighting matrix is no longer positive definite, departing fundamentally from the traditional LQ problem. By introducing a backward-stochastic-differential-equation (BSDE) in such MV frameworks, one can accommodate indefinite and even negative definite control weights under certain conditions. Optimality of the BSDE design was established by embedding the original problem into a tractable auxiliary problem.

To take into consideration random environments not representable via the usual stochastic differential equation setup, we developed more precise models with possible random switching in regimes [14]; based on these results, we have studied the following models for platoon control.

Switching Diffusion Model At the cyber level, each vehicle is a node in the network of n_0 vehicles. The node state $x_i(t)$ consists of the vehicle's position, speed, direction, etc. The ith vehicle's dynamics can be described by

$$dx_i(t) = f_i(x_i(t), u_i(t), \alpha(t))\,dt + \sigma_i(x_i(t), u_i(t), \alpha(t))\,dw_i, \ \ i = 1, \ldots, n_0 \ \alpha(t) \in M,$$

where the first and second terms represent vehicle dynamics and noise effects, respectively; $w_i(\cdot)$ is a standard Brownian motion; u_i is the control for the ith vehicle. The switching network topology G is a Markov chain $\alpha(\cdot)$, taking values in $M = \{1, \ldots, m_0\}$ with generator Q which is independent of the Brownian motion $w(\cdot) = (w_1(\cdot), \ldots, w_{n_0}(\cdot))^T$. At time t, a vehicle uses the available neighborhood information from the graph $G(\alpha(t))$ to adjust its control $u_i(t)$. For each $\alpha(t) \in M$, the drift $f_i(x_i(t), u_i(t), \alpha(t)) : R \times R \mapsto R$ delineates the "average behavior" of the dynamics, and $\sigma_i(x_i(t), u_i(t), \alpha(t)) : R \times R \mapsto R$ tells us the "standard deviation" of the dynamics about its mean. Acting as the intensity of the noise, σ_i is referred to as *diffusion coefficient* in probability.

If σ_i is large, the system displays a wide range of fluctuations. If the noise is sufficiently small, the system is represented by a dynamic system with only the drift term. When the random disturbances disappear completely, the system reduces to a deterministic system that does not include the Brownian motion part. Then, the interconnected systems can be represented by a hybrid stochastic system

$$dx(t) = f(x(t), u(t), \alpha(t))\,dt + \sigma(x(t), u(t), \alpha(t))\,dw,$$

where $x(t) = (x_1(t), \ldots, x_{n_0}(t))^T \in R^{n_0}$, $u(t) = (u_1(t), \ldots, u_{n_0}(t))^T \in R^{n_0}$, w is a n_0-dimensional standard Brownian motion, $f(x, u, \alpha) = (f(x_i, u_i, \alpha)) \in R^{n_0}$, $\sigma(x, u, \alpha) = \text{diag}(\sigma_i(x_i, u_i, \alpha)) \in R^{n_0 \times n_0}$. Since the Markov chain $\alpha(\cdot)$ takes values in a set of isolated points, it is a discrete event process. Thus the system framework is a hybrid switching diffusion, in which continuous vehicle dynamics and discrete topology switching events coexist. To date, such switching models has only been used to treat networked platoon control in our work [11], especially when delays are involved. In addition, we can also study a switching jump diffusion model as proposed in our recent work. The switching jump diffusion model is similar to (1) but with an additional Poisson jump term $k(\cdot)$, which is used to model short noise and sudden burst of noise not representable by the usual Brownian motions.

MDP Formulation At a given instant t, if $\alpha(t) = i$, then $G(\alpha(t)) = G(i)$, namely the topology switches according to the values of $\alpha(t)$. Given $\alpha(t) = i$, the dynamics of the platoon are given by a controlled Markov chain x(t) instead of a differential equation. x(t) takes values in R^{n_0}, which is another continuous-time Markov chain with general state space $M_1 \subset R^{n_0}$ and generator Q(u, i) that is control dependent. This controlled Markov chain is a Markov decision process. Once the state i is fixed, its dynamics are completely determined by the generator. This formulation differs from the standard setup of MDPs in that the generator Q is not only u-dependent, but also α-dependent (i.e., $Q = Q(u, \alpha)$). Thus, we write it as

$$x \simeq Q(u, \alpha), \alpha \in M, M_1 \subset R^{n_0}$$

Here, the state space may be uncountable or a subset of R^{n_0}, which is more general than the MDP formulations with either a finite or countable state space.

These two formulations have their own pros and cons. The switching diffusion model contains detailed descriptions of the system dynamics and its solution is associated to certain differential equations. The Markov decision process uses distributed information in which the controlled Markov chain has control-dependent generators, leading to a simplified model structure.

Mean-variance control for utility-safety management With dynamics described by either the switching diffusion (1) or by the Markov decision process (2), we can define the mean-variance platoon distribution control problem. Motivated by our recent work, our main objective is to achieve platoon formation in a short horizon, in which the associated dynamic programming equation is time dependent. Let $U(\cdot) : R^{n_0} \mapsto R$ be a sufficiently smooth and concave utility function based on factors such as inter-vehicle distances. Our objective is to find an admissible strategy $u(\cdot)$ among all the admissible actions whose expected terminal value is $E\,U(x(T)) = U(z)$ for some given $z \in R^{n_0}$, and the risk as measured by the variance of the terminal wealth (i.e., $\text{Var}U(x(T)) = E[U(x(T)) - EU(x(T))]^2 = E[U(x(T)) - U(z)]^2$) is minimized.

Then the mean-variance problem is a constrained stochastic optimization problem, parameterized by $z \in R$:

$$\begin{cases} J(x_0, i_0, u(\cdot), \lambda) = E|U(x(T)) - U(z)|^2, \\ \text{subject to } EU(x(T)) = U(z), \ (x(\cdot), u(\cdot)) \text{ admissible} \end{cases}$$

The problem is called feasible if there is at least one strategy satisfying the constraint. It is said to be finite if it is feasible and the infimum of $J(x_0, i_0, u(\cdot))$ is finite. An optimal strategy to the above problem, if it exists, is said to be an efficient strategy corresponding to $U(z)$, and the corresponding $Var(U(x(T)), U(z))$ and $(\sigma_{U(x(T))}, U(z))$ are interchangeably called an efficient point, where $\sigma_{U(x(T))}$ denotes the standard deviation of $U(x(T))$. The set of all the efficient points is called the efficient frontier.

4 Software-Defined CAV Infrastructure

As new CAV technologies (e.g., CAV communication with predictable reliability [2, 3]) are developed, we need to evaluate these technologies in real-world settings before the enabled new CAV applications can be deployed for real-world use, thus requiring a methodology for enabling the spiral innovation process of research, development, pilot deployment, and adoption. CAV is inherently an interdisciplinary field involving multiple disciplines such as transportation, communication, and control, it is thus important for different disciplines to collaborate in the innovation process for the concerted progress in all dimensions of CAV technologies [15].

As shown in Fig. 4a, vehicle technologies have traditionally been researched, developed, and evaluated using a few test vehicles, and then the technologies can be deployed in real-world settings at scale since the individual vehicles are controlled separately without communicating with one another. With CAVs, however, this innovation paradigm has inherent drawbacks since the testing with a few vehicles in

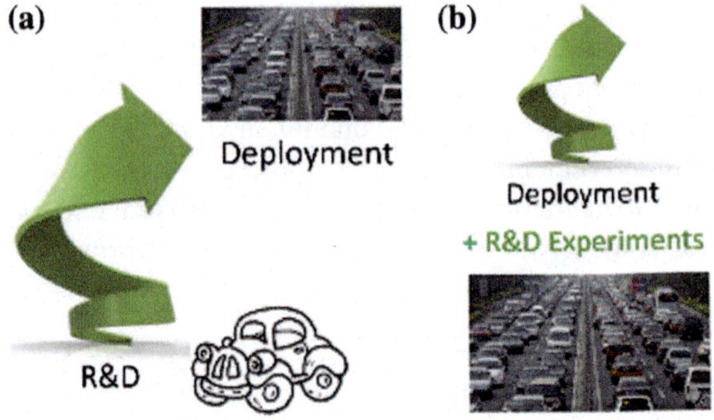

Fig. 4 Vehicle innovation paradigms

selected test settings does not capture CAV communication behavior and its impact on CAV safety, efficiency, and comfort in real-world deployment of a large number of vehicles in potentially unpredictable traffic, road, and environmental conditions. To address this issue, the USDOT Safety Pilot Model Deployment of connected vehicles (CVs) and the planned large scale CV pilot deployments employ thousands of test vehicles in real-world settings. While these deployments enable transportation research such as understanding the benefits of CVs for the safety and efficiency of road transportation, these deployments do not enable the testing of technologies in other important dimensions of CAVs, for instance, the testing of new CAV networking and control technologies. Accordingly, these deployments cannot serve as experimental infrastructures for the research communities of different CAV disciplines to collaborate with one another. For continuous, cross-disciplinary innovation and evolution of CAV technologies, there still lacks methodologies that enable a symbiotic innovation paradigm where, as shown in Fig. 4b, research experiments from multiple CAV disciplines may co-exist with one another and with existing CAV technologies and applications in real-world traffic [15].

Towards enabling the symbiotic innovation paradigm for continuous, cross-disciplinary CAV evolution, we develop a software-defined CAV infrastructure through software-defined virtualization (SDV) and multi-domain emulation (MDE) of CAVs. SDV partitions each CAV sensing, computing, and networking equipment into multiple "slices" such that the sensing, computing, and networking tasks of a CAV experiment and those of the existing CAV applications can execute in different slices without interfering with one another, thus enabling the use of vehicles in real-world traffic to support CAV experiments. SDV dynamically allocates sensing, computing, and networking resources among CAV experiments and existing applications according to the requirements of CAV experiments and applications, and, by allowing the research communities of different CAV disciplines to share a same experimental infrastructure, SDV enables cross-discipline collaboration and coordination in the spiral CAV innovation process. By deploying and executing the "field" components of CAV emulation as CAV experiments in vehicles, SDV also enables integrating vehicles in real-world traffic with multi-domain simulation of V2X communication, vehicle dynamics, and traffic flow in high-performance cloud computing infrastructures, thus integrating the high-fidelity of real-world vehicle traffic with the flexibility and scalability of in-cloud simulation. To leverage existing accomplishments from different CAV communities and to facilitate cross-discipline collaboration, our multi-domain emulation (MDE) system develops mechanisms for integrating the state-of-the-art network simulator ns-3 and vehicle as well as traffic dynamics simulator SUMO into a holistic CAV emulation system where different CAV communities can contribute to different parts of the emulation system and leverage capabilities developed by other communities [15].

To demonstrate our software-defined CAV infrastructure and to evaluate its effectiveness, we study CAV technologies and applications in public safety which has not been well explored as an application domain of CAVs. In particular, we deploy our SDV-based CAV platforms in the police patrol vehicles of Wayne State

University. Together with GENI networking and cloud computing infrastructures, these police patrol vehicles effectively support the real-world application of 3D-vision-based public safety surveillance and MDE-based CAV experiments at the same time, thus enabling symbiotic experiments of emerging CAV technologies and applications in real-world vehicle traffic of existing CAV technologies and applications for public safety, which we expect to speed up the evolution and adoption of CAV technologies and applications [16].

5 Concluding Remarks

In this chapter, we have reviewed the challenges that cyber-physical dynamics and uncertainties pose to CAV wireless networking, sensing, and control, and we have proposed an integrated, cross-layer framework for establishing a trustworthy foundation for CAVs. To realize the integrated framework, we have made initial progress towards an algorithmic and methodological foundation for CAV wireless networking and control. For enabling continuous CAV evolution, we have also developed a software-defined CAV infrastructure for symbiotic CAV experiments and real-world deployments as well as for cross-discipline collaboration in CAV innovation. These algorithms, methodologies, and software-defined infrastructures enable the development and deployment of CAV solutions such as those for active safety, networked fuel economy optimization, and platooning in general. Interesting future work include, but are not limited to, extending the PRKS wireless transmission scheduling algorithm to address vehicle mobility in CAV networks, joint optimization of CAV networking, sensing, and control, and integrating the open-source driving simulator OpenDS into our software-defined CAV infrastructure.

References

1. Johri R, Rao J, Yu H, Zhang H (2015) A multi-scale spatiotemporal perspective of connected and automated vehicles: applications and wireless networking. arXiv:1508.05344
2. Zhang H, Che X, Liu X, Ju X (2014) Adaptive instantiation of the protocol interference model in wireless networked sensing and control. ACM Trans Sensor Netw 10(2)
3. Zhang H, Liu X, Li C, Chen Y, Che X, Lin F, Wang LY, Yin G (2015) Scheduling with predictable link reliability for wireless networked control. In: IEEE/ACM IWQoS
4. Hellerstein J, Diao Y, Parekh S, Tilbury DM (2004) Feedback control of computing systems. Wiley-IEEE Press
5. Liu X, Chen Y, Zhang H (2015) A maximal concurrency and low latency distributed scheduling protocol for wireless sensor networks. Int J Distrib Sensor Netw (Hindawi)
6. Che X, Zhang H, Ju X (2014) The case for addressing the ordering effect in interference-limited wireless scheduling. IEEE Trans Wireless Commun

7. Tobagi F, Kleinrock L (1975) Packet switching in radio channels: Part II–the hidden terminal problem in carrier sense multiple-access and the busy-tone solution. IEEE Trans Commun COM-23(12)
8. Shladover S, Desoer C, Hedrick J, Tomizuka M, Walrand J, Zhang W, McMahon D, Peng H, Sheikholeslam S, McKeown N (1991) Automated vehicle control developments in the PATH program. IEEE Trans Vehicular Tech 40:114–130
9. Li SE, Zheng Y, Li K, Wang J (2015) An overview of vehicular platoon control under the four-component framework. In: IEEE intelligent vehicles symposium, 2015
10. Li SE, Zheng Y, Li K, Wang J (2015) Scalability limitation of homogeneous vehicular platoon under undirected information flow topology and constant spacing policy. In: Chinese control conference, 2015
11. Yang Z, Yin G, Wang YL, Zhang H (2013) Near-optimal mean-variance controls under two-time-scale formulations and applications. Stochastics, 723–741
12. Markowitz H (1952) Portfolio selection. J Finance 77–91
13. Zhou XY, Li D (2000) Continuous-time mean-variance portfolio selection: a stochastic LQ framework. Appl Math Optim 19–33
14. Zhou XY, Yin G (2003) Markovitz's mean-variance portfolio selection with regime switching: a continuous time model. SIAM J Control Optim 1466–1482
15. Wang Y, Jin H, Li C, Zhang H, Hua J, Rao J, Riley G, Holt A, Gossman P (2015) Symbiotic CAV Evolution: software-defined infrastructure and case study in public safety (working paper)
16. Wang Y, Jin H, Li C, Zhang H, Hua J (2015) CAV applications and networks: wireless networked 3D mapping for public safety. https://youtu.be/y_QxXA0MJzI

Enabling Technologies for Vehicle Automation

Mohammed Yousuf, Daniel J. Dailey, Sudharson Sundararajan and Ram Kandarpa

Abstract Vehicle automation relies heavily on technologies such as sensing, wireless communications, localization, mapping, human factors, and several others. Applications planned within the USDOT's automation research roadmap depend on the understanding and applicability of these technologies. Thus it is important to be aware of the state of these technologies, and more importantly to stay ahead of the curve. The value of this task is not in accurately predicting the future of these technologies for USDOT's automation program, but to minimize surprises. A four step process was followed to better understand advances in positioning, navigation and timing (PNT), mapping, communications, sensing and human factors. The first step identified the needs, second tracked high-level trends and based on these findings, the third step identified gaps. Finally, these insights were used to develop potential next steps for USDOT consideration. Paper presents a high-level overview of the research process, findings from the study and insights on next steps.

Keywords Automated vehicles · Enabling technologies · Localization · Mapping · Self-driving vehicles · Sensing · Human factors

---

M. Yousuf (✉) · D.J. Dailey
Turner-Fairbank Highway Research Center, 6300 Georgetown Pike,
Mclean, VA 22101, USA
e-mail: Mohammed.Yousuf@dot.gov

D.J. Dailey
e-mail: Daniel.Dailey.ctr@dot.gov

S. Sundararajan · R. Kandarpa
Booz Allen Hamilton Inc., 20 M Street, Suite 1000, Washington D.C. 20003, USA
e-mail: Sundararajan_Sudharson@bah.com

R. Kandarpa
e-mail: Kandarpa_Ram@bah.com

G. Meyer and S. Beiker (eds.), *Road Vehicle Automation 3*,
Lecture Notes in Mobility, DOI 10.1007/978-3-319-40503-2_18

1 Background

Today both traditional and non-traditional auto companies are investing heavily in research projects to develop automated applications for vehicles (traffic jam assist, automated braking etc.) that minimize the role of the driver. Increasingly, a car's capabilities are determined more by its electronics and software than by its mechanics. The industry is envisioning a system of vehicle-to-vehicle communications, cameras, variety of sensors (Radar, LIDAR, RFID, etc.) and other devices integrated with advanced algorithms that can monitor the road in a variety of roadway, weather and traffic conditions to enable driverless systems.

Vehicle automation can be divided into the following major subsystems that implement a control loop, as described in automated vehicle reference model architectures such as the National Institute of Standards and Technology/4D Real-time Control System (NIST 4D/RCS) architecture:

1. Sensing—acquire raw data from vehicle sensors
2. Perception—process the raw sensor data and create usable information on the vehicle and its environment
3. Planning—use the perception of the vehicle's current environment and situation as well as a priori data to plan short and long term actions and behaviors
4. Control—implement the short and long term actions and behavior.

This high level control loop is only intended to illustrate common approaches for many automated vehicles to date. This control loop also shows an example for conceptual processes and components of each subsystem. It is not intended to be an all-inclusive list of sensors, algorithms and behaviors, and in fact specific hardware and software approaches can vary wildly across the industry. The control loop in Fig. 1 mirrors the approach that we as humans take in sensing, perceiving, and interacting with our world.

To implement these subsystems, a wide variety of well-entrenched and emerging technologies are integrated to provide the appropriate level of vehicle automation. These enabling technologies are grouped into the following high-level categories and were researched as part of this research effort (Fig. 2).

2 Research Approach

The research team scanned a combination of approximately 250 popular and technical press sources, investigating over 100 key observations to extract needs, trends and gaps in each technology area.

Based on these insights next steps for USDOT considerations were developed. A graphic representing this high-level process is provided in Fig. 3 for clarity. In addition to exploring the current state of technology, the research team also developed predictions about near- (5 years), mid- (10 years), and long-term

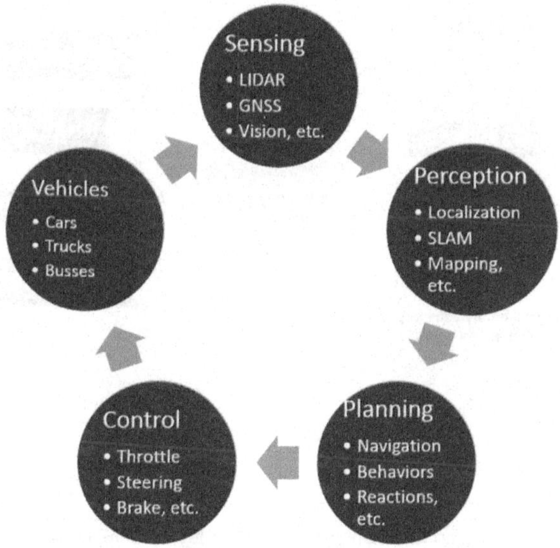

Fig. 1 High-level control loop

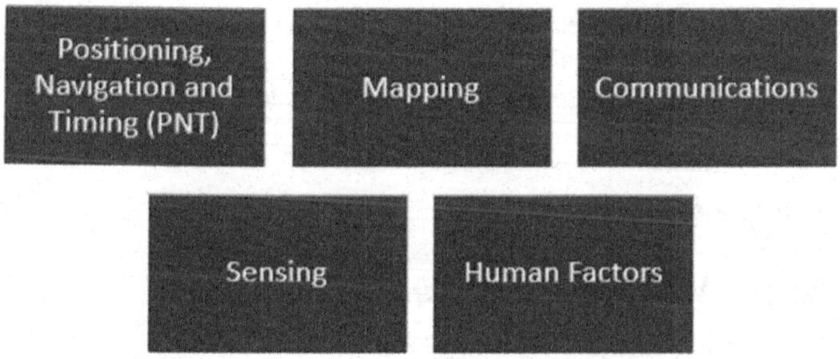

Fig. 2 Research categories

(15 years) technology milestones. Through an analysis of the team's observations, trends were extrapolated to illustrate possible future technology advances that will enable higher levels of vehicle automation. These predictions are designed to help USDOT better understand the role they can play in shaping and accelerating the deployment of automated vehicle features.

Different levels of automation are enabled by advances in the above mentioned technologies. The NHTSA [1] defines different levels of automation as given in Table 1.

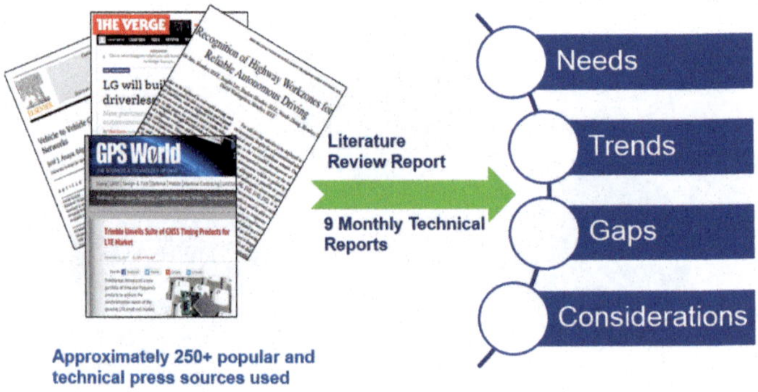

Fig. 3 High level process

Table 1 Levels of automation (NHTSA)

Automation	Description
Level 0	Absolutely no automation (Referred as legacy vehicle in this paper)
Level 1	Involves one or more independent systems assisting the driver to drive the vehicle. E.g.: Electronic Stability Control assists driver to regain control
Level 2	At least two primary controls of the vehicle work in unison. E.g.: adaptive cruise control with lane centering
Level 3	Vehicle drives itself under normal and expected conditions; however, the driver takes control when required
Level 4	These can drive occupied or unoccupied as long as it has a destination input

3 Summary of Findings

3.1 Position, Navigation and Timing

Positioning, Navigation, and Timing (PNT) technologies require high accuracy and reliability to enable vehicle automation. To meet these needs, researchers and product manufacturers are continuing to develop positioning systems that are smaller, faster and much more capable than their predecessors and are at much lower price points. These positioning systems are evolving to not just provide an output based on a combination of Global Navigation Satellite System/Inertial Navigation System (GNSS/INS) but also to provide frameworks that can incorporate other data such as wheel speed encoder odometry, visual odometry, and data from other sensors. In parallel, redundant techniques are emerging to provide precise timing in situations where GNSS is unavailable. This capability could greatly help automated vehicle applications that are dependent upon precise time synchronization for cooperative aspects.

To address the gaps that still remain for PNT, advancing research to provide corrections to vehicles from roadside infrastructure and how that applies to vehicle automation has high potential. Enhanced navigation information could be provided from the infrastructure as part of the digital infrastructure initiative. This information could include detours and safe harbors for automated vehicles in scenarios where the technology decides to intervene with the human driver. Investigating redundant timing techniques and approaches for safety critical key infrastructure components is also critical to advance PNT to enable automated vehicle applications. Furthermore cyber security is a major concern but many do not consider the vulnerabilities of GNSS signals. More research is needed for this area as well.

3.2 Mapping

Mapping technologies are focused mainly on macro-level maps that would allow a vehicle to navigate along corridors, cities, etc. as well as mapping the immediate area around a vehicle to determine its relative position and orientation with respect to everything around it. These algorithms must operate in real-time and be able to handle challenging conditions and dynamic environments. At the macro level, private sector companies are continuing to collect and develop extremely high resolution maps that could be used by vehicle automation systems ranging from Advanced Driver Assistance Systems (ADAS) to fully automated, driverless vehicles. In parallel, researchers are continuing to advance Simultaneous Localization and Mapping (SLAM) techniques to improve the capabilities of visual SLAM in challenging lighting conditions (dawn, dusk, night, rain, fog) as well as changing environments (seasonal changes, cities). Researchers have also been developing techniques for cooperative vehicle mapping to allow one vehicle to utilize the perception capabilities of another vehicle to more safely navigate through its environment. Finally the increasing availability and performance of Graphical Processing Units (GPU) is enabling these complex algorithms to run in real-time.

While advancements are continually being made in mapping technologies, there are significant gaps that still remain. Information that is being proposed to be provided from the infrastructure is limited in nature (e.g. SAE J2735 SPAT/MAP/TIM). This could be significantly enhanced by sharing a more rich set of situational awareness information between static infrastructure points and vehicles to enable vehicle automation. Research in this area is conducted to investigate reliable and efficient methods of exchanging situational awareness between infrastructure points and vehicles which could greatly improve safety of vehicles (at all levels of automation) around intersections, work zones, school zones, etc. A richer set of map data to be shared from the infrastructure can also be very useful. Finally, building off of research that was conducted on sensor-friendly infrastructure markings that take into consideration many of the new vehicle perception and localization techniques will help advance mapping to the next stage.

3.3 Communications

The successful implementation of connected/automated vehicles depends on a reliable and secure communication infrastructure. Recent advances in communications, controls, and embedded systems are transforming vehicles to sensor platforms, collecting data from other cars and the cloud, and feeding it back to drivers and infrastructure to assist in safe navigation, pollution control, and traffic management. The next step in this evolution is utilizing Machine to Machine (M2M) communication and Internet of Things (IoT) and creating the Internet of Autonomous Vehicles. The promise of 5th generation wireless systems (5G) with its unique capability of transmitting gigabits of data will pave the way for this evolution. Autonomous vehicles rely on spectrum for high-performance radar, laser, LIDAR (light detection and ranging) systems, and satellite communications for location mapping. Given the tight spectrum resource, it is important to consider alternative technologies for autonomous vehicles including cellular networks, 4G/5G, Wi-Fi, and VLC/Li-Fi among others.

Expansion of M2M and telematics in the vehicular networks exposes the network to external cyber-security threats. As such ensuring vehicular network security is an important part of these developments. Security areas that need attention include: security of the infrastructure, encryption of all the communications on the cloud, and building robust authentication schemes. Also, as smart devices are integrating with vehicles there is an added threat of personal information theft that should be addressed. Moreover successful implementation of connected/automated vehicles requires infrastructure investment, coordination among international standard bodies, and enabling policies, laws, and regulations development.

3.4 Sensing

The need for advanced in-vehicle technologies and intelligent infrastructure constantly drive innovations in sensor technology. The new sensor technologies are either enhancing the current sensor capabilities to have a better range, accuracy, and robustness or to develop completely new smarter and smaller sensors. Overall, the focus of sensor technology has been on generating quality data that supports development of new safety applications and capabilities. These improved sensors not only provide accurate data but also provide data at a higher frequency. As a result the market has also seen advances in microprocessors that manage sensors and analyze sensor data to provide meaningful outputs for control systems. These microprocessors enable sensor fusion to provide redundancy for safety applications, making sensors more powerful. Sensor fusion thus improves efficiency and enables the development of several automated vehicle safety applications. The use of cloud based systems to analyze real-time data from a variety of sensors and other dynamic events is also gaining popularity and further enables sensor fusion. Technologies which process data in real-time provide opportunities to develop several safety applications.

Accurate and reliable 360° sensing of vehicle surroundings, sensor fusion and data accessibility, enhanced performance in adverse weather conditions, and more robust standards and safety mandates for V2V and V2I applications are some of the needs and gaps identified. Though most of the sensor related technologies will be vetted by the market place, tracking these advances could be of potential importance for infrastructure applications that support automated vehicle applications. For example using stereo cameras at a complex urban intersection to get a 360° view of the intersection could help develop better signal timing strategies.

3.5 Human Factors

A high level of trust must exist between the user and the technology to enable vehicle automation dependency. Consequently, the human factor research of automated vehicles is vital for the acceptance/usage of such a technology. Based on the recent technology observations, there is a need for cooperative and non-distracting devices to support the different level of automation and the smooth transition from human control to the automated control. In addition, the mixed environment of automated vehicles and legacy vehicles still need to be investigated. Some of the observed trends based on the conducted research showed that the driver state sensing technology to mitigate driver distraction is underway by several OEMs. It was also found that many driver information/assistance display technologies are claimed to be ready for market usage. Although the automated vehicles technologies are rapidly evolving, there are many research gaps that still remain. For example, the research related to interaction of automated vehicles with other vulnerable road users (also pedestrians, cyclists, and legacy vehicles) is still lacking. Another example, is that, the impact of many of the current technologies (including after-market devices) related to HMI on safety is still unknown; especially for the different levels of automation.

Though many of the current technologies (including after-market devices) related to Human Machine Interface (HMI) may potentially enhance safety of legacy vehicles and improve driver awareness, their impact on safety is still doubtful. At the end, it can be stated that there is a need to investigate the influence of this type of technologies on enabling vehicle automation on the streets; especially when it comes to elderly drivers or drivers with disabilities and/or special needs. This kind of analysis is very critical in the development of a framework to assess safety of self-driving algorithms. Safety for pedestrians and bicyclists also can be assessed using surrogate safety studies that can model autonomous vehicles explicitly.

4 Discussion

Based on research findings, a high-level summary of how different technologies map to different levels of automation is depicted in Fig. 4. These technologies are seen as cumulative, i.e., for communications, the advent of 5G vehicle automation applications does not replace DSRC. This figure shows the progression of technologies relative to other technologies in the same technology area, and does not imply that technologies that are vertically aligned will occur at the same time. It is critical to note that there are many technologies and combination that can enable different levels of automation, e.g., to solve the problem of high accuracy self-localization, one application may rely more on sensor fusion while another may rely on more expensive inertial measurement units. Therefore this figure represents only those technologies that were encountered in this research process.

The research observations were used as the basis for predicting which technologies will most profoundly impact vehicle automation in 5, 10, and 15 years. These key technologies, described in Fig. 5, will be instrumental in achieving the high level of reliability and safety that different automated vehicle applications require.

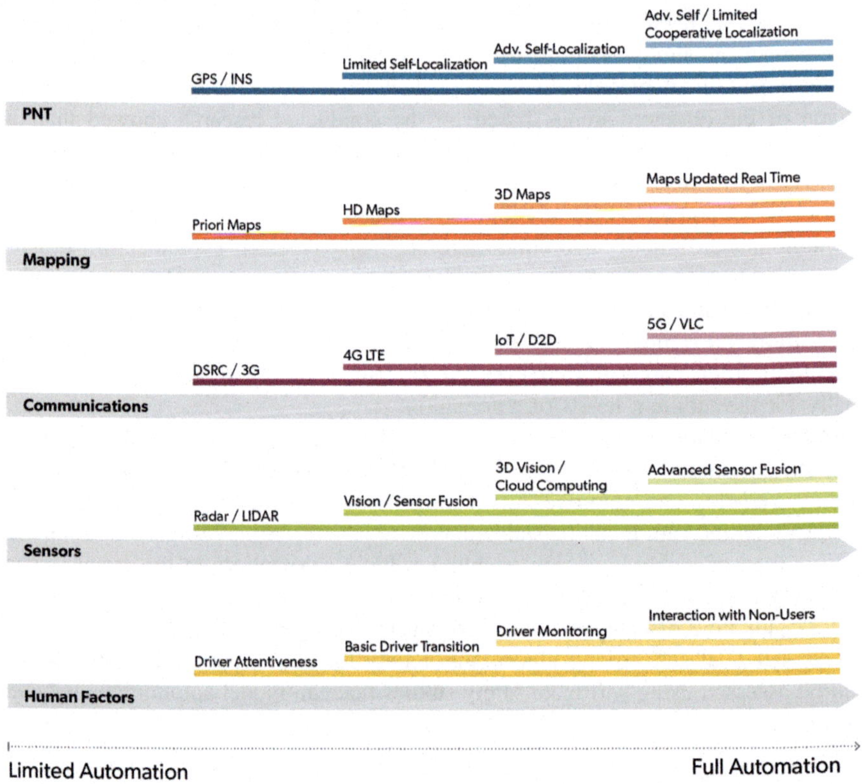

Fig. 4 Advances in key technologies will enable higher levels of automation

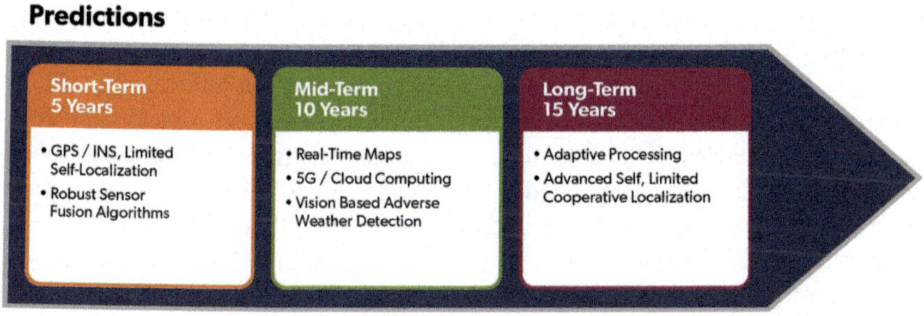

Fig. 5 Predictions for near, mid, and long term advances in technology that will profoundly impact and advance automated vehicles

Significant improvements in self-localization are expected due to contributions from GPS/INS, SLAM, maps, infrastructure support, and cooperative localization. Self-localization can currently be estimated within about 1.5 m in ideal conditions. Within 5 years, advances in GPS, INS, and SLAM [2] with higher definition maps [3] will enable "limited self-localization" to within about 10 cm. To achieve the next level of accuracy within 10 years (approximately 2 cm accuracy), better SLAM and real-time maps [4] will be needed that capture dynamic changes in the environment. Within 15 years, further advances in GPS, infrastructure support (e.g., RF localization like UWB), and cooperative localization [5] will allow for "advanced self-localization" to within 1 cm, which is critical for Level 3 and 4 automation.

Algorithms are an area where considerable advances are expected. Stereo vision cameras have been recently used more as a less expensive alternative to improve the reliability of sensors and provide a wider range of detection for different ADAS applications. Currently fully autonomous vehicles are using a combination of LIDARs, radars, and vision based sensing systems to complement the weaknesses of each kind and provide redundancy [6]. Based on the observations the team made during the course of this research, there is a push towards vision based sensing as the potential first class player in the area of perception technology [2]. The need for sensor fusions to combine the data elements received from multiple sensors in order to improve performance and reliability of the overall system has also been focused on. Advances in sensor fusions are needed to address challenges in the higher levels of automation, such as robust object detection and classification, adverse weather conditions, etc. In particular AI techniques, such as machine learning, are proving to be effective at improving processing speed and performance [7]. AI will be crucial to achieving Level 4 automation, and is predicted to reach a high level of functionality within 15 years.

Many companies are exploring the use of cloud data sharing and computing and potentials to wirelessly connect cars, cyclists, and pedestrians to the cloud. Using cloud data along with advanced sensing technologies may potentially improve the detection, warning, and auto braking stages of collision avoidance applications.

Also by analyzing the data in real-time and updating the map on the cloud, information such as adverse environmental conditions [8], traffic jams, accidents, and other dynamic events can be systematically transmitted to the vehicle so it can adjust its route or other course of actions [9]. Low latency, high bandwidth communication is needed to enable this information sharing. Current advances in communication are occurring rapidly [10], which gives the research team confidence that 5G will be delivering value to automation within 10 years.

The complex interaction between human and machine will continue to be a challenge for years to come, especially as infotainment continues to compete for driver's attention [11]. Achieving safe Level 3 operation will require a better understanding of transition of control between driver and machine, which require simulators and test beds that are just now coming online [12]. As with automation in the aviation industry, challenges related to transition of control will likely continue to persist, perhaps beyond the horizon of predictions made in this report. Driver monitoring technologies are still being researched [13], and could begin to play a significant role in automation in the next 10 years. The research team found research on the impact of automated vehicle technology and other road users sparse [14], and expects this challenge will persist until much higher market penetration of automated vehicle technologies and aftermarket solutions [15] are achieved.

Acknowledgments The authors gratefully acknowledge the input of the panel speakers and participants during the special session of *Enabling Technologies for Vehicle Automation* as part of the 2015 Automated Vehicles Symposium. The authors are also grateful to the researchers involved in this study (Sara Sarkhili, Mahsa Ettefagh, Ismail Zohdy, Shawn Kimmel, and Patrick Chuang from Booz Allen Hamilton; and Mike Brown from Southwest Research Institute).

References

1. NHTSA Website – Press Releases. http://www.nhtsa.gov/About+NHTSA/Press+Releases/U.S.+Department+of+Transportation+Releases+Policy+on+Automated+Vehicle+Development. Accessed 8 Oct 2014
2. Milford M, Vig E, Scheirer W, Cox D (2014) Vision-based simultaneous localization and mapping in changing outdoor environments. J Field Robot 31(5)
3. Avdhut J, James MR (2015) Toyota research institute, North America, generation of accurate lane-level maps from coarse prior maps and lidar. IEEE Intell Transp Syst Mag
4. Auto News http://www.autonews.com/article/20141117/OEM06/311179997/heres-looking-at-you—in-3-d. Accessed 17 Nov 2014
5. Li H, Tsukada M, Nashashibi F, Parent M (2014) Multivehicle cooperative local mapping: a methodology based on occupancy grid map merging. IEEE Trans Intell Transport Syst 15(5)
6. ITS International: Drivers want semi-autonomous safety features. http://www.itsinternational.com/categories/location-based-systems/news/drivers-want-semi-autonomous-safety-features/. Accessed 7 Nov 2014
7. Feroz MGI, Jada C, Kumar R, Yenala H (2015) Map based representation of navigation information for robust machines learning. Sign Process Commun Eng Syst (SPACES)
8. Gallen et al (2015) Nighttime visibility analysis and estimation method in the presence of dense fog. IEEE Trans ITS 16(1)

9. Driving: Volvo wants to eliminate car crashes using the Internet. http://driving.ca/volvo/auto-news/news/volvo-wants-to-eliminate-car-crashes-with-the-internet. Accessed 7 Jan 2015
10. Qualcomm: Qualcomm Drives Future of Automotive Connectivity with New 4G LTE Modems. https://www.qualcomm.com/news/releases/2015/03/02/qualcomm-drives-future-automotive-connectivity-new-4g-lte-modems#
11. Forbes, Can Panasonic's 'e-Cockpit' make you a better driver? http://www.forbes.com/sites/joannmuller/2014/12/24/video-see-how-panasonic-wants-to-make-you-a-better-driver/#1dfd42a7715a. Accessed 24 Dec 2014
12. ITS International: TRL's DigiCar driving simulator becomes fully automated. http://www.itsinternational.com/categories/location-based-systems/news/trls-digicar-driving-simulator-becomes-fully-automated/. Accessed 18 Nov 2014
13. Bigthink, Car Sensors Monitor Drivers for Distraction, Tiredness. http://bigthink.com/ideafeed/new-car-sensors-monitor-drivers-for-distraction-tiredness, 2015
14. Gizmag, smrtGRiPS could smarten up your handlebars. http://www.gizmag.com/smrtgrips-smart-handlebar-grips/35571/. Accessed 12 Jan 2015
15. Gizmag, CarVi brings modern driver assist technology to older cars. http://www.gizmag.com/carvi-driver-assist-retrofit/35985/. Accessed 8 Feb 2015

Technical Evaluation and Impact Assessment of Automated Driving

Felix Fahrenkrog, Christian Rösener, Adrian Zlocki and Lutz Eckstein

Abstract Currently different research activities on automated driving are conducted around the globe. The European flagship research project on automated driving functions is the research project "AdaptIVe" (Automated Driving Applications and Technologies for Intelligent Vehicles). Besides the development of automated driving functions, the project deals with general research on legal aspects, human factors and evaluation. The evaluation and impact assessment of automated driving functions faces different challenges considering the complexity of the technology. In this context, this paper describes the evaluation approaches that are taken in the project for the technical evaluation and impact assessment.

Keywords AdaptIVe · Evaluation · Technical assessment · Impact assessment

1 Introduction

Automated driving has been a vision since the early 20th century. A first step towards vehicle automation was the introduction of cruise control in the Chrysler Imperial as the so called "auto pilot" in 1958 [1]. A first step towards this vision was the introduction of ADAS (advanced driver assistance systems) in the last

F. Fahrenkrog (✉) · C. Rösener · L. Eckstein
Institut für Kraftfahrzeuge, RWTH Aachen University,
Steinbachstraße 7, Aachen, Germany
e-mail: fahrenkrog@ika.rwth-aachen.de

C. Rösener
e-mail: roesener@ika.rwth-aachen.de

L. Eckstein
e-mail: eckstein@ika.rwth-aachen.de

A. Zlocki
fka Forschungsgesellschaft Kraftfahrwesen mbH Aachen,
Steinbachstraße 7, Aachen, Germany
e-mail: zlocki@fka.de

G. Meyer and S. Beiker (eds.), *Road Vehicle Automation 3*,
Lecture Notes in Mobility, DOI 10.1007/978-3-319-40503-2_19

decade of the 20th century. Following the successful introduction of ADAS research on higher automated driving functions has been ongoing since many years. These functions were intensively investigated and demonstrated during the DARPA Challenges [2, 3] as well as activities of Google and their so called Google self-driving cars [4] in the US and in Europe by the Berta Benz Drive [5] and the GCDC [6].

This chosen path is continued by the European research project AdaptIVe, which develops new automated driving applications in order to promote safer and more efficient driving [7]. Within AdaptIVe automated driving functions are developed for different speed ranges (parking, urban and highway driving) and at different automation levels [8] ranging from level 2 up to level 4 functions.

Due to its nature as a research project, AdaptIVe focuses less on the market introduction of automated driving, but rather on contribution to fundamental research questions of automated driving. These are the legal framework for automated driving functions [9], the human interaction with these functions [10] and the evaluation. For the latter one the objective is to set up a general evaluation framework that can be applied for different types of automated driving functions including a safety impact analysis and an environmental impact analysis.

2 Evaluation Methodology

The initial point for the definition of an evaluation framework for automated driving functions in AdaptIVe was a review of existing evaluation approaches for ADAS and automated driving functions.

During the function development typically a continuous and iterative technical evaluation is conducted. The main objective of this evaluation is to check, whether the pre-defined requirements are fulfilled by the functions and whether the defined performance is reached (e.g. SARTRE [11], HAVEit [12] or KONVOI [13]). This type of evaluation can be seen in the sense of a verification process, which is defined according to [14] as the evaluation of whether or not a product, service, or system complies with a regulation, requirement, specification, or imposed condition. It is often a company- or project-internal process.

On the other hand a technical evaluation can be conducted more in the sense of a validation, which is defined as "the assurance that a product, service, or system meets the needs of the customer and other identified stakeholders" [14]. Here, the function is assessed against certain pre-defined evaluation criteria. In this context important evaluation criteria are the acceptance and, in particular for a research project, the impact on traffic. This has been investigated in detail for different ADAS functions (e.g. TRACE [15], interactIVe [16], eIMPACT [17]), but not for automated driving functions. A general evaluation methodology for this evaluation stage has been introduced by the PReVAL [18] project. This approach considers three evaluation areas (technical, user-related and safety impact assessment).

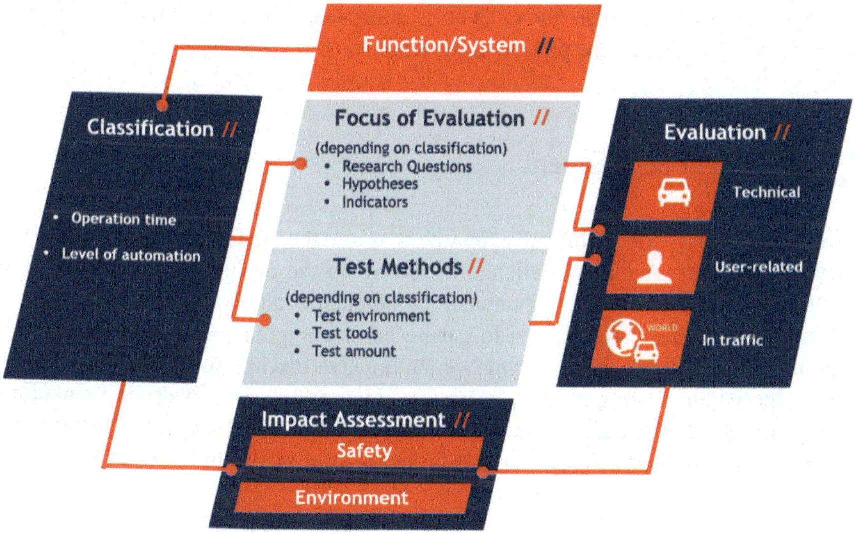

Fig. 1 Evaluation framework in AdaptIVe

The AdaptIVe evaluation framework (see Fig. 1) takes up the PReVAL approach of three different evaluation areas and extents it by a fourth evaluation area that is analysing the changes in the traffic behaviour of the surrounding traffic. The following evaluation areas are considered in AdaptIVe: technical assessment, user-related assessment, in-traffic assessment and impact assessment (safety and environmental effects). For AdaptIVe the focus is mainly on the validation of the developed functions, however also the verification is considered in the technical assessment. Within the AdaptIVe evaluation framework for each evaluation area the tools and indicators that should be applied are described in detail [19]. In general similar tools are used as for the evaluation of ADAS (field test, test in controlled environment, simulator test and simulation). However, the applied tools need to be adapted to the tested automated driving function. Therefore, different automated driving functions have been reviewed. The results of this review is that next to the automation level the operation time must to be considered when selecting appropriate evaluation tools [6].

Regarding the operation time functions have to be distinguished:

- Functions that operate only for a short period of time (seconds up to few minutes). Typical examples are automated parking functions and the minimum risk manoeuvre function. These functions are called *event-based* operating functions.
- Functions that once they are active can be operated over a longer period of time (minutes up to hours). A typical example for this type of functions is a highway pilot or a motorway automation function. These functions are called *continuous operating* functions.

In the following chapters the evaluation approaches for the technical assessment and impact assessment are presented in detail.

3 Technical Assessment

The focus within the technical assessment is on the performance of automated driving functions. A major challenge in setting up the evaluation framework is to limit the test effort in order to guarantee an efficient evaluation while ensuring that all important aspects are covered to ensure an as comprehensive as possible evaluation. Since the operation regime of automated driving functions is covering nearly the whole driving process, a large number and variety of driving situations needs to be taken into account [20]. It is obvious that detailed analyses of all driving situations it is not feasible due to the limited resources, as shown in Fig. 2.

3.1 Technical Assessment for Event-Based Operating Functions

For the assessment of event-based operating functions an approach similar to the assessment of today's ADAS functions is used (e.g. PReVAL [18], interactIVe [21]). According to this approach the first step the assessment scope by means of the research questions is defined. Based on these questions, related hypotheses are

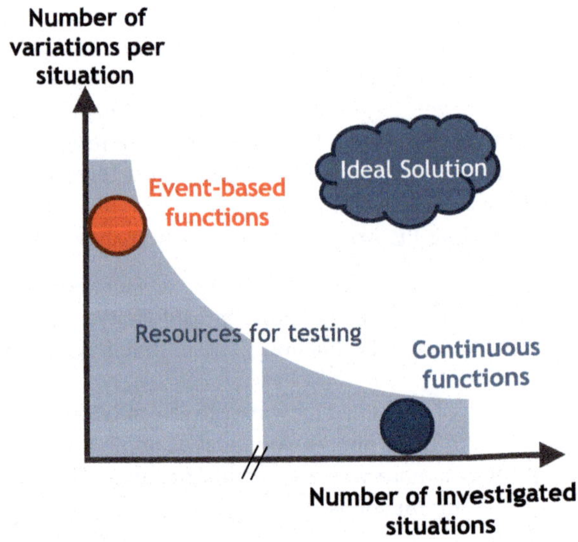

Fig. 2 Test approach for the technical assessment within AdaptIVe

defined to be analysed. Therefore adequate performance indicators are chosen for the analysis.

Once the definition of the assessment scope is defined the relevant test cases are selected. The basis for the test cases is typically the function's use cases respectively other situations that are relevant like e.g. accident scenarios. The actual testing is the second last step of this approach. The tests are typically conducted in a controlled field—mainly test tracks or in closed test garage for parking scenarios. During the testing the parameter set of the test case (e.g. initial velocities or relative distances) is varied. Ideally, each test—a test case with certain parameter set—is repeated several times in order to ensure statistically valid results. The analysis of the test data is the last step of the process. This step includes the calculation of derived measures as well as indicators. Based on the indicators the pre-defined hypotheses are analysed.

3.2 *Technical Assessment for Continuously Operating Functions*

For the continuously operating functions the focus is slightly different in the technical assessment. Since these functions cover different driving situations also for the assessment, a wider scope in the assessment is required. For the continuously operating functions the focus is less on the performance in a certain driving situation but rather on overall performance during the whole driving process.

Therefore, it is not useful to define certain single test cases. Instead a holistic assessment approach that covers as many different driving situations as possible is needed. Such an assessment approach is the (small) field test on public roads, in which the function must be able to handle different driving situations. The drawbacks of the field test approach are both, the rather uncontrolled test set-up and the relatively high effort for a field test in general. Hence, the extent of the field test needs to be limited to a feasible amount.

Analogue to event-based functions, the assessment approach of continuously operating functions starts with the definition of the research questions and hypotheses. This includes the definition of adequate performance indicators and criteria for the assessment of the hypotheses as well. In order to investigate the performance over the whole driving process adequate indicators are required. For this purpose the basic requirements for automated driving functions need to be considered, which are:

- The function must drive the vehicle in safe manner,
- The function must be able to operate in mixed traffic conditions,
- The function should not affect the other traffic in a negative way.

These basic requirements imply that the automated driving functions need to operate within the range of normal driving behaviour and should at least be as safe as non-automated driving.

Thus, the baseline for the assessment should be the human driver respectively his/her behaviour. Since the driving behaviour of each human driver is different, it can only be described in distributions. These distributions of driver behaviour need to be obtained before the actual assessment is performed. For this purpose two approaches are used in AdaptIVe:

- Data of previous field test projects will be used (e.g. filed operational tests (FOT) like the euroFOT project [22]), since it provides information on the driving behaviour of many different drivers.
- Each test route is driven several times before the test with and without the function in order to consider specific characteristics of the test region.

Next to the distribution of normal human driving also legal boundaries must be considered for the assessment (e.g. speed limits, restriction on passing) respectively their violation.

In the next step the actual tests are prepared. This includes the definition of the test route and test length. To limit the test effort to a feasible extend, the test route has to be chosen in a way that all relevant driving situations will occur with a certain probability. Therefore, the required test length respectively duration needs to be estimated a priori based on the number of expected relevant driving situations that occur while driving in public traffic. For this estimation, the data and the knowledge gained in previous field operational tests is used.

The actual tests are split into two steps. The first step is the pre-tests, in which the basic functionality is checked as well as the accuracy of the used sensors is analysed. These tests are conducted similar to the event-based functions tests on a test track. If the tests have been finished and the function is operating properly, the main test on public roads will be conducted. During the tests the test route will be driven several times with and without the function under assessment. After the tests, the data is evaluated analogue to the process described for the event-based function. The analysis will be conducted on the level of the whole test trip as well as on the level of certain relevant driving situations.

In order to enable a holistic technical assessment of the automated driving function in all relevant driving situations, the previously described methodology foresees a situation-based assessment. During the tests on public roads different driving situations occur stochastically. For identification and classification of these driving situations, detection-rules-based [23] or machine-learning-based approaches [24] can be used. These approaches are used to identify the following situations: Free driving, car following, lane change (left/right), cut-in of other vehicle (left/right) and approaching object.

In order to ensure that all relevant driving situations are occurring during the test, the length of the small field test is estimated by means of data from previous FOT, such as euroFOT [22]. The reference data of the field operational test is clustered in

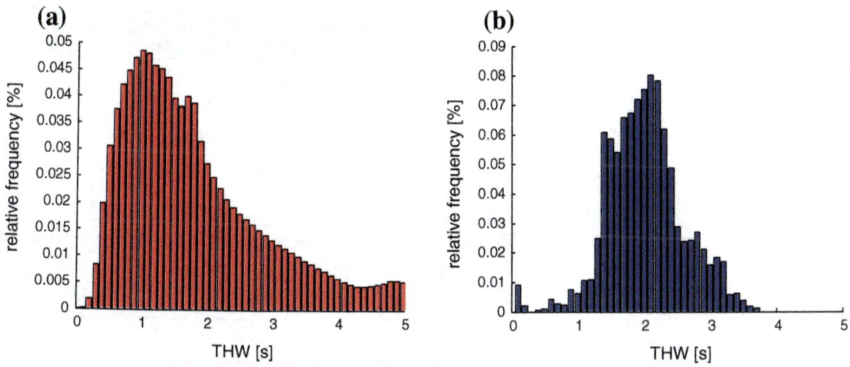

Fig. 3 Distribution of indicator time headway within the scenario "car following" for reference/euroFOT (*left*) and for the exemplary test of an automated function (*right*)

relevant driving situations by using a situation space approach. Afterwards, the distribution of spatial frequencies of all relevant driving situations is calculated. In accordance to [20] a Poisson distribution for the occurrence of driving situations is assumed and the minimal test length for the occurrence of at least k = 5 driving situations is calculated. While most driving situations are detected by using data obtained from field operational tests, driving situations influenced by the infrastructure, e.g. "enter motorway" are obtained by using statistics regarding the infrastructure of motorways in the test region.

After classification of the relevant driving situations the predefined hypotheses can be evaluated. For this purpose, it will be checked, whether the distribution of the performance indicator of the tested automated driving function deviates from the distribution of the reference data. The distributions of the indicator time headway (THW) within "car following" situations are shown for reference/euroFOT (left) and with an automated driving function for the exemplary test (right) in Fig. 3.

Driving with the automated driving function shifts the distribution of the performance indicator THW to become compared to the driving without the automated driving function. The median value of the THW increases from 1.5 s in the reference data to 2.0 s in the tests with the function. The analysis of the consequences of this effect with respect to the traffic in terms of efficiency and safety is part of the impact assessment.

4 Safety Impact Assessment

In the safety impact assessment it should be analyzed, which benefits in terms of avoided accidents or injuries automated driving functions can provide. Different approaches (field of application analysis, accident re-simulation, and field test) were developed and applied for ADAS in the past [25]. However, for the impact

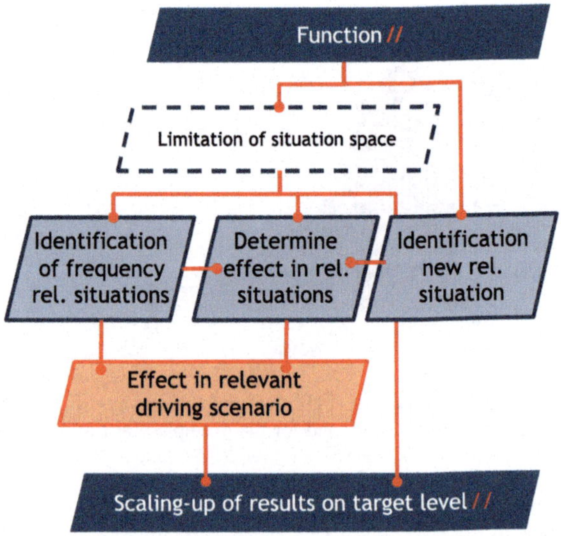

Fig. 4 Safety impact assessment approach in AdaptIVe

assessment of automated driving function different challenges compared to the impact assessment for ADAS must be considered:

- Today's accident data do not consider collisions of automated vehicles,
- Automated driving function operate already before a critical situation occurs, which makes the re-simulation of accidents more difficult
- Interaction with other road users (automated/non-automated) in mixed traffic condition has only rarely been studied so far.

In order to overcome these challenges an approach for the safety impact assessment has been defined that bases on [26] and consists of three main steps (see Fig. 4):

1. *Identify relevant situations*
   By means of the first step all relevant (critical) situations should be identified as well as their frequencies. Therefore, by means of microscopic traffic simulation the traffic flow at different road section is simulated. The microscopic traffic simulations allow to vary the penetration rates and to identify changes in the frequencies of driving situations depending on the penetration rate.
2. *Investigate the relevant situation in detail*
   In this step the effect of functions is investigated. The applied method is similar to the re-simulation. This means that certain driving situations are simulated with and without the function under study. The effect of the function is determined by comparison of both results. The analyzed driving situations include accidents as well as critical driving situation. In this step also the results of the other assessment (technical, user-related, in-traffic assessment) are considered.

3. *Identification of new situations*
   Automated driving function will cause also (critical) situations that are not part of today's traffic. Examples are the "transition of control" or "minimum risk maneuver" situation. These situations are investigated in the third step by means of risk analysis

5 Summary and Outlook

The paper describes the evaluation approach for the technical and safety impact assessment that is taken in the European research project AdaptIVe. A major challenge is the diversity of automated driving functions—ranging from automated parking function to automated highway driving functions at different automations levels. Therefore, in the developed evaluation framework the applied evaluation tools are chosen based on the operation time of the function. The developed evaluation framework is applied in AdaptIVe.

The evaluation approach for AdaptIVe does not consider the context of market introduction relevant validation/verification of sufficient safety operation of automated driving functions after the market introduction. To tackle this issue ika has already introduced the circle of relevant situation approach [27, 28]. This approach bases on two basic ideas, which are

- a combination of existing test tools in effective and cost-effective manner,
- re-usage of logged field data to cover by means of simulation the overall situation space.

Thus, the aim is to extend the AdaptIVe framework with the approaches that focus on the safety validation for the field in order to get to a comprehensive evaluation framework for automated driving functions.

Acknowledgments The research leading to these results has received funding from the European Commission Seventh Framework Program under the project AdaptIVe, grant agreement number 610428. The authors would like to thank all partners within AdaptIVe for their cooperation and valuable contribution. In particular the authors want to thank the partners of the Evaluation SP, namely BAST, BMW, CTAG, CRF, TNO and Lund University.

References

1. Rowsome F (1958) What it's like to drive an auto-pilot car. Popular Science Monthly, USA
2. NN Report to Congress, DARPA Prize Authority, DARPA, 2006
3. NN DARPA Urban challenge Media. DARPA urban challenge, http://archive.darpa.mil/grandchallenge/, 2014
4. Markoff J (2010) Google cars drive themselves. The New York Times

5. Ziegler J, Bender P, Schreiber M et al (2014) Making Bertha drive – an autonomous journey on a historic route. IEEE Intell Transp Syst Mag 6(2)
6. Ploeg J, Shladover S, Nijmeijer H et al (2012) Introduction to the special issue on the 2011 grand cooperative driving challenge. IEEE Trans Intell Transp Syst 13(3)
7. Etemad A, Bartels A (2014) A stepwise market introduction of automated driving. In: 21st ITS World Congress, Detroit
8. NN Taxonomy and Definitions for Terms Related to On-Road Automated Motor Vehicles, SAE document J3016, 2014
9. Knapp A (2015) Legal issues addressed in the EU funded AdaptIVe project. In: 22nd ITS World Congress, Bordeaux, 2015
10. Johansson E (2014) Human factors in vehicle automation - activities in the European project AdaptIVe. VRA Webinar
11. Dávila A (2012) Validation test plan and results. SARTRE, Deliverable D4.1 & 4.2
12. Beutner A, Pettersson S, Seglö F et al (2011) Optimized and validated demonstration vehicles. HAVEit, Deliverable D13.1
13. Deutchle S (2006) Das KONVOI-Projekt – Entwicklung und Untersuchung des Einsatzes von elektischen LKW-Konvois auf Autobahnen, 15. Aachner Kolloquium
14. NN Adoption of the Project Management Institute (PMI®) Standard A Guide to the Project Management Body of Knowledge (PMBOK® Guide)–4th edn. IEEE Guide, vol 2011
15. Karabatsou V, Pappas M, Elslande P et al (2006) A-priori evaluation of safety functions effectiveness – Methodologies, Trace, Deliverable D4.1.3
16. Fahrenkrog F, Zlocki A, Dobberstein J et al (2013) Impact assessment of developed application – overall interactive assessment, interactIVe, deliverable D7.5
17. Wilmink I, Janssen W, Jonkers E et al (2008) Impact assessment of intelligent vehicle safety systems. eIMPACT, Deliverable D4
18. Scholliers J, Hendirks F, Aust ML et al (2008) Project final report and recommendation for future assessments. PReVAL project, Deliverable DL IP D12/D16.4
19. Rodarius C, Duflis J, Fahrenkrog F, et al (2015) Test and evaluation plan. AdaptIVe Deliverable D7.1
20. Wachenfeld W, Winner H (2015) Die Freigabe des autonomen Fahrens, Autonomes Fahren, Technische, rechtliche und gesellschaftliche Aspekte. Springer Vieweg
21. Larsson P, Esberg I, van Noort M (2012) Test and evaluation plans. interactIVe Deliverable D7.4
22. Kessler C, Etemad A, Alessandretti G (2012) Final Report, euroFOT, Deliverable D11.3
23. Benmimoun M, Fahrenkrog F, Zlocki A, Eckstein L (2012) Detection and classification of critical events by means of vehicle data. ATZ Worldwide 114(10):60–66
24. Reichel M, Botsch M, Rauschecker R, Siedersberger KH et al (2010) Situation aspect modeling and classification using a scenario based random forest algorithm for convoy merging situations. In: 13th International IEEE Conference on Intelligent Transportation Systems, 2010, pp 360–366
25. Page Y, Fahrenkrog F, Fiorentino A et al (2015) A comprehensive and harmonized method for assessing the effectiveness of advanced driver assistance systems by virtual simulation: the PEARS initiative. In: 24th international technical conference on the enhanced safety of vehicles (ESV)
26. Fahrenkrog F, Zlocki A, Eckstein L (2014) Bewertung aktive Sicherheitssysteme vom Test zur Wirksamkeitsanalyse. Automobiltechnische Zeitschrift 01
27. Eckstein L, Zlocki A (2013) Safety potential of ADAS – combined methods for an effective evaluation, 23rd ESV
28. Zlocki A, Eckstein L, Fahrenkrog F (2015) Evaluation and sign-off methodology for automated vehicle systems based on relevant driving situations. In: 94th Annual TRB Meeting, Washington D.C.

Part V
Transportation Infrastructure and Planning

Integrated Traffic Flow Models and Analysis for Automated Vehicles

Bart van Arem, Montasir M. Abbas, Xiaopeng Li, Larry Head, Xuesong Zhou, Danjue Chen, Robert Bertini, Stephen P. Mattingly, Haizhong Wang and Gabor Orosz

Abstract With the emergence of connected and automated vehicle (CAV) technologies, research on traffic flow modeling and analysis will play a very important role in improving our understanding of the fundamental characteristics of traffic flow. The frontier of studies on CAV systems have examined the impacts of CAVs on freeway bottleneck capacity, and macroscopic traffic flow, CAV

B. van Arem
Delft University of Technology, 2600 GA Delft, Room 4.13, Delft, The Netherlands
e-mail: b.vanarem@tudelft.nl

M.M. Abbas
Virginia Polytechnic Institute and State University, 750 Drillfield Drive, 200 Patton Hall, Room 301-D3, Blacksburg, VA 24061, USA
e-mail: abbas@vt.edu

X. Li
University of South Florida, 4202 E. Fowler Avenue, ENG 207, Tampa, FL 33620-5350, USA
e-mail: xiaopengli@usf.edu

L. Head
The University of Arizona, 1127 E. James E. Rogers Way, Room 111, P.O. Box 210020, Tucson, AZ 85721-0020, USA
e-mail: larry@sie.arizona.edu

X. Zhou
Arizona State University, College Avenue Commons, Room 474, 660 S. College Avenue, Tempe, AZ 85287-530, USA
e-mail: xzhou74@asu.edu

D. Chen (✉)
University of Wisconsin-Madison, 1415 Engineering Dr., Room 2205, Madison, WI 53703, USA
e-mail: danjuechen@gmail.com

R. Bertini
California Polytechnic State University San Luis Obispo, 1 Grand Avenue, San Luis Obispo, CA 93407-0353, USA
e-mail: rbertini@calpoly.edu

G. Meyer and S. Beiker (eds.), *Road Vehicle Automation 3*, Lecture Notes in Mobility, DOI 10.1007/978-3-319-40503-2_20

applications on optimization of individual vehicle trajectories, potentials of CAV in traffic signal control, and applications of CAV in network routing. For current and future research initiatives, the greatest challenge lies in the potential inconsistencies between user, operator, and manufacturer goals. Specific research needs were identified on data collection and analysis on CAV behavior and applications. This paper summarizes the presentations and discussions during the Automated Vehicles Symposium 2015 (AVS15) held in Ypsilanti, Michigan, on July 20–23, 2015.

Keywords Traffic flow model · CAV behavior · Data collection · Research needs

1 Introduction

As connected and automated vehicle (CAV) technologies emerge in road transportation systems, it is critical to bring together the wireless communication, vehicle dynamics and traffic flow communities to better understand the fundamental characteristics of traffic flow with varying level of connectivity and automation and identify research needs to develop models to assess the implications of CAV in various aspects, including mobility, safety, and the environment.

This paper summarizes the presentations and discussions and presentations during the breakout session on Integrated Traffic Flow Models and Analysis for Automated Vehicles at the Automated Vehicles Symposium 2015 (AVS15) held in Ypsilanti, Michigan, on July 20–23, 2015. In the breakout section five scholars were invited to present their frontier research on CAV from different perspectives. Following the presentations, a panel consisting of the five invited speakers had extensive discussions with the audience to identify the key challenges in this research area define and further research needs.

The remaining sections of this paper are organized as follows: Sect. 2 presents the summary of the invited presentations and Sect. 3 introduces the key results from the panel discussion. References are also provided at the end.

S.P. Mattingly
University of Texas at Arlington, 425 Nedderman Hall, 416 Yates St., Box 19308, Arlington, TX 76019, USA
e-mail: mattingly@uta.edu

H. Wang
Oregon State University, Owen Hall, Room 307, Corvallis, OR 97331, USA
e-mail: haizhong.wang@oregonstate.edu

G. Orosz
G034 Autolab, University of Michigan, 1231 Beal Avenue, Ann Arbor, MI 48109-2121, USA
e-mail: orosz@umich.edu

2 Frontier Research on CAV Traffic Flow Modeling and Control

This section presents a summary of the invited talks, which addressed the research frontier from various aspects. The summary includes the background and significance of the research, the main conclusions, and also directions for future research.

2.1 *Modeling Traffic Flow with Automated Vehicles at Highway Bottlenecks*[1]

Automated vehicles are believed to help resolve congestion in motorway networks. Back of the envelope pipeline capacity calculations based on short distance platooning indeed easily yield capacity estimates up to 6,000 veh/h, which by far exceeds traditional capacity estimates that are around 2200 veh/h as can be found in the Highway Capacity Manual [1]. As a consequence it has been argued that automated vehicles could triple roadway capacity. Shladover et al. [2] have shown lane capacities to increase from (approximately) 2,065–3,970 veh/h using more realistic models for the behaviour of manual and cooperative adaptive cruise control (CACC) equipped vehicles. In real freeway networks, however, congestion typically occurs at bottleneck locations such as lane drops, on-ramps, weaving sections or other discontinuities in the road geometry. Bertini and Hoogendoorn [3] argue that congestion can be resolved by improving outflow from and reducing inflow into road segments, preventing spill backs and spreading traffic over different alternatives.

van Driel and van Arem [4] focused on the impacts of a congestion assistant on a high traffic demand motorway stretch with a 4–3 lane drop scenario. The congestion assistant was assumed to help a driver to slow down by an active gas pedal when approaching congestion and provided full longitudinal automation in congested traffic. Using traffic flow simulations, they have shown that around 30 % delay saving could be achieved at low CAV penetration rates such as 10 %. The delay savings are particularly caused by increased outflow of the congested area. The free flow bottleneck roadway capacity did not change significantly.

van Arem et al. [5] studied a special lane where vehicles equipped with CACC can drive at short intra platoon distances, again at a motorway stretch with a 4–3 lane drop. The motivation to study a special lane for CACC platoon was to explore whether this could help reduce congestion at a high traffic demand level. The problem was studied using traffic flow simulation. The experiments showed again that increased outflow results in higher throughput. Notably, it was also found that the mandatory lane changes of manual vehicles moving out of the CACC-only lane

---

[1]By Bart van Arem, Delft University of Technology (Netherlands).

caused detrimental impacts on traffic efficiency. Therefore, the geometric design involving a dedicated lane only beneficial at a specific range of CACC penetration.

The results of the studies outlined above and others [6] suggest that the contribution of automated vehicles to reducing traffic congestion depends substantially on how automated vehicles perform at bottleneck locations compared with human drivers. The current literature also suggests that aiming to improve outflow of congestion by automated vehicles (and reducing capacity drop) can yield early benefits at low penetration rates. In addition, there is a need to develop coordination strategies of automated vehicles in complex road geometries such as weaving areas and to manage transitions between manual and automated driving and between general purpose and dedicated lanes. A recently started project conducted by UC Berkeley PATH and TU Delft on behalf of the Federal Highway Administration (FHWA) [7] is focusing on active platoon formation strategies, supported by road side systems and dedicated lanes. The project includes traffic flow simulations of a 21-km SR-99 freeway stretch with high occupancy vehicle (HOV) lanes near Sacramento.

2.2 *Impacts of Connected and Automated Vehicles on the Macroscopic Fundamental Diagram*[2]

The fundamental diagram (FD) (Fig. 1) shows the relationships between traffic flow, speed, and density on freeways and arterial links. Traditionally, these relationships have been used in traffic modeling, control, and operation assessments. The fundamental diagram can be used for queue estimation and control (e.g., platoon progression) at signalized intersections, and it defines the main building blocks of traffic flow theory. The FD is primarily used on freeways to model the steady-state relationship between density and flow [8]. Temporary changes in traffic flow and/or capacity on a link usually result in shockwaves where conditions change rapidly. Collecting data in such dynamic situations typically results in a cloud of data points, and consequently, a non-unique FD.

Recent developments in traffic flow and network modeling gave rise to what is now widely known as the macroscopic fundamental diagram (MFD). The MFD has been defined as a relationship between average network flow and accumulation. This concept attempts to establish an FD-like relationship at the network level. It requires averaging of all network flows and density over small time periods, and plotting their relationship [9, 10] for the purpose of establishing large-scale network state estimation and control [9]. Recent work indicated that the level of heterogeneity in the network characteristics affects the degree of data scatter and maximum flow of an MFD. Researchers recommended constructing the MFD for

[2]By Montasir M. Abbas, Virginia Tech.

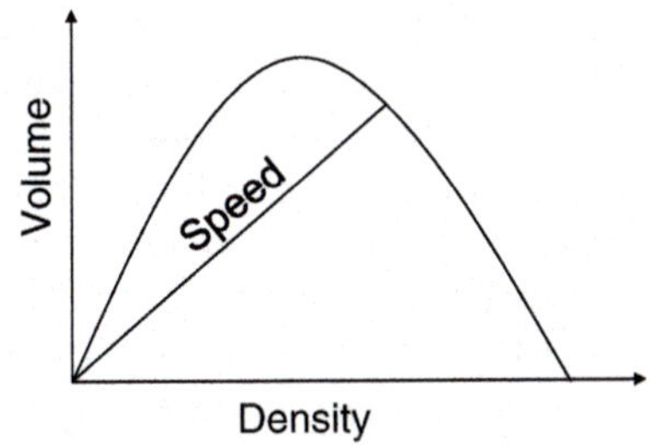

Fig. 1 Fundamental diagram

homogeneous networks with low link density variance to reduce the scatter [11, 12].

The introduction of connected and automated vehicles into the traffic stream is bound to change the shape and form of the fundamental diagram. To model future connected and automated vehicles, a novel multi-agent reservation system was developed where vehicles reserve space tiles and time slots at signalized intersections in a priority-stratified first-come first-served way (Fig. 2). However, higher priority vehicles can revoke the reservation of lower priority vehicles, forcing them to re-reserve later time slots.

In contrast to the MFD, a new concept was defined that aims to explore the heterogeneity of the FD relationship associated with each priority level. This may allow one to separate FDs associated with different priorities from. This concept is called aggregated fundamental diagram (AFD) and aim to define a class of service (COS) for each data set. Vehicles with priorities that fall onto a higher capacity AFD experience a higher COS. One could expect links that suffer from spillback, low capacities, or suboptimal control to fall onto lower AFD and lower COS.

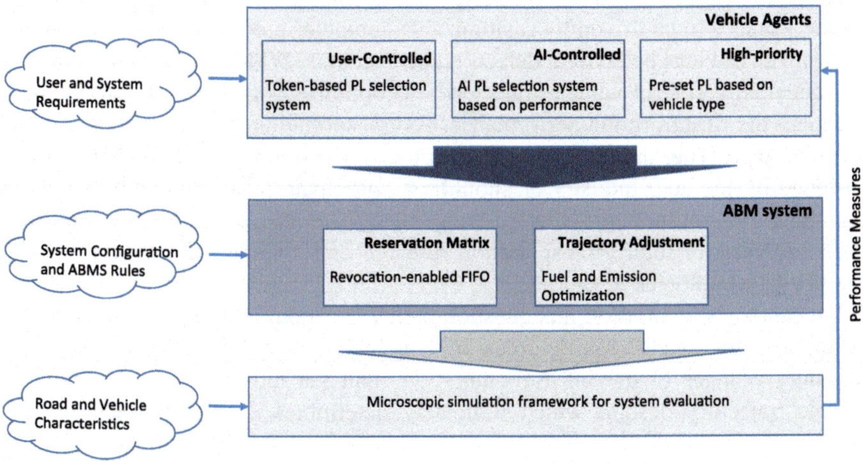

Fig. 2 Reservation multi-agent system (MAS)

A future objective may be to observe and analyze the impact of the introduction of high priority connected and automated vehicles on the fundamental diagram and develop control applications.

The proposed agent-based framework was implemented using the VISSIM traffic simulation software [13]. An experiment was conducted for modeling agents' self-organization ability by setting different priorities for agents according to their associated movement phases. Through movements on the main approaches were given the highest priorities, followed by through movements on the minor approaches, then the left turners. Conflict resolution was handled by the MAS reservation system (Fig. 2) and without signalized intersection operation.

No models were fitted to the data to avoid extrapolation issues, but it is clear that different priority levels are falling onto different FDs. It is therefore concluded that:

- AFDs can be used for characterization and improvement of network performance.
- Agent-based modeling and simulation provides a suitable framework for analysis of automated and connected vehicles.
- Movements with different priorities exhibit different traffic flow relationships. The framework illustrated here removes the necessity to assign higher priorities (and hence a higher classes of service) to approaches. This allows the redefinition of "major" and "minor" movements to align with individual user's needs and urgency levels regardless of the approach/phase they are in.

2.3 *Automated Vehicle Trajectory Control Versus Classical Traffic Flow Theories*[3]

One can anticipate that future connected and automated vehicle technologies will enable vehicles to be driven by customizable computer programs instead of human beings with obvious behavioral defects and constraints. With these technologies, we can potentially control vehicle trajectories to optimize both the individual experience and the overall traffic performance. As transportation researchers, do we just focus on modifying the current infrastructure to accommodate automated vehicles designed by the auto industry or should we participate in the research agenda of designing and optimizing vehicle driving rules? The answer to this question may shift the scope of future transportation research and change the impacts of these emerging technologies to our society.

The author's opinion on this question is that we should participate in designing or modifying vehicle driving rules in conjunction with thinking about corresponding changes of the infrastructure. We shall get into the very depth of the vehicle trajectory design, which ultimately determines each individual driver's

[3]By Xiaopeng Li, University of South Florida.

experience as well as the collective traffic performance. Only by allowing infrastructure interacting with vehicle trajectory control, can we maximize the overall benefit of these emerging technologies while minimizing each individual's cost on transportation.

As one of the pioneering attempts, a novel shooting heuristic (SH) was proposed for optimizing CAV trajectories, considering realistic constraints including vehicle kinematic limits, traffic arrival patterns, and car-following safety [14]. In particular, the relationships between SH results and those obtained from classic traffic theories were investigated. The time geography theory was generalized to investigate the theoretical properties of SH. It was found that SH can be viewed as a generalization of a number of classic traffic flow models (e.g., kinematic wave model and Newell's simplified car-following model). Numerical examples were conducted to illustrate these theoretical findings. This study reveals commonalities and connections between emerging trajectory optimization approaches and classic traffic flow theories and it lays a solid foundation for devising holistic control strategies for future transportation systems with emerging technologies.

Three directions are worth investigation for future research: (a) vehicle trajectory optimization, (b) next generation infrastructure design, and (c) interaction between automated vehicles and human drivers.

2.4 *Automated and Connected Vehicles on Traffic Signal Control*[4]

The contribution examines the question of what is new in a connected and automated environment for traffic signal control. Traffic signal control evolves from the efficiency-oriented design to the collective fashion considering multi-model efficiency/safety as well as quality of service. There seem to be enormous application potentials, particularly in four major aspects:

1. Reducing the start-up lost time using signal phase and timing data (SPaT) information and cooperative adaptive cruise control (CACC) control.
2. Reducing the change period lost time by replacing yellow interval with green clearance interval and eliminating red clearance. This can be achieved by using SPaT data to accurately compute the Go or No Go decision for every vehicle.
3. Reducing the saturation headway of CACC vehicle to increase the utilization of green time (saturation flow period).
4. Increasing system capacity by forming self-organized platoons (e.g., using CACC technology).

There are some key research questions in this area that need to be addressed, including (a) what's what are the requirements for CACC on signalized roadways?

[4]By Larry Head, University of Arizona.

(b) How to improve traffic signal efficiency under different market penetrations? And (c) how to provide incentives to form self-organizing platoons?

2.5 *Connected and Automated Vehicles on Space-Phase-Time (SPT) Hyper-Network*[5]

Emerging automated vehicle technology technologies will allow for route guidance for individual vehicles and have the potential for close coordination between vehicles and the infrastructure. In practice, these possible benefits also create new challenges. As an example, in urban areas, it is critical to provide individual vehicles with guided paths and optimize related traffic signals together so as to maximize both traffic safety and mobility. This problem is referred to as simultaneous route guidance and traffic signal optimization problem (RGTSO). In this contribution, the RGTSO problem is was formulated using the time-dependent space-time network coupled with a new type of network traffic signal control representation, referred to as phase-time network. The new formulation of RGTSO can guarantee the RGTSO problem's linearity. The RGTSO problem is then decomposed into two sub-problems: the route guidance (RG) problem and the traffic signal optimization (TSO) problem. The solution to the TSO sub-problem provides time-dependent link capacity constraints for the RG sub-problem whereas the dual prices of the RG sub-problem indicate search directions for the TSO sub-problem. Both the RG and TSO sub-problems can be solved using a computationally efficient finite-horizon dynamic programming framework, enhanced by scalable parallel computing techniques. Two numerical experiments demonstrated that the system optimum of the RGTSO problem can be quickly reached with relatively small duality gap for medium-size urban networks.

3 Discussion

The panel discussion (including audience interaction) identified the key challenges in traffic flow research in the connected-automated environment and outlined the future research needs, which not only help to advance research on traffic flow modeling of CAV but also to promote the collaboration and coordination of the traffic flow research community with other communities, including vehicle automation, cyber-physical systems, and human factors.

It was concluded that the *biggest challenge* lies in the potential inconsistency in user, operator, and manufacturer goals. In particular, users are most likely interested in benefits at the individual level, such as improving safety, reducing travel time,

[5]By Xuesong Zhou, Arizona State University.

and enhancing travel convenience and comfort. In contrast, traffic and transportation system operators mostly prioritize the system-wide benefits. Finally, manufacturers are indeed driven by vehicle sales. Therefore, the interests of these key stakeholders are not always well aligned, which could severely impede the technology implementation and application.

For example, traffic operators are quite interested in the appealing benefits of CACC vehicles to improve roadway capacity and mitigate congestion (i.e., positive societal benefits), which require drivers to form short-space platoons. As a user, however, there is no obvious benefit to motivate such platooning maneuver. From the manufacturer's perspective, user acceptance is a key factor to decide further investment. Therefore, it's critical to understand and align the interests of the different stakeholders. The research community can play a key role in this process. For example, control/management strategies can be designed that will benefit both users and the system and thus connect the user, operator, and manufacturers. To enable this, the research community needs to understand the operational opportunities and challenges of CAV, which requires operational data, most specifically operational constraints of CAV. This will need the cooperation particularly from the manufacturers. In addition, it is important to recognize that the introduction of CAV technologies will occur incrementally, so for the foreseeable future there will always be a mix of vehicle types and technologies on our roadways. Research aimed at assessing the potential benefits of a range of vehicle types is also important.

Based on the discussion, future research needs can be classified into two groups.

1. Data collection and analysis. It's important to collect data and study (a) changes in driver behavior (especially in non-automated vehicles), (b) CAV operational capabilities and constraints, (c) interactions of drivers with CAV capabilities, and (d) benefits of CAV to consumers (especially linked to applications). Detailed trajectory data are needed.
2. CAV applications. Research is needed to (a) analyze the traffic impacts of CAV on corridor and network level operations under various market penetration rates, and (b) investigate trajectory control and vehicle cooperation strategies at freeway bottlenecks (e.g., merge and weave bottlenecks) and traffic signals, especially under multiple objectives/constraints (e.g., safety, environment, time, comfort, driver acceptance).

It was agreed by the breakout session participants that traffic flow related research plays a critical role in advancing and implementing the CAV technologies, and that collaboration with other research communities, such as vehicle automation and cyber-physical systems, will be very beneficial.

References

1. Transportation Research Board, Highway Capacity Manual 2010, Transportation Research Board, 2010
2. Shladover SE, Su D, Lu X (2012) Highway capacity increases from cooperative adaptive cruise control. In: Proceeding of ITS World Congress, 2012
3. Hoogendoorn SP, Bertini R (2012) Can we control traffic? Instilling a proactive traffic management culture. Delft University of Technology, Essencia The Hague
4. van Driel CJG, van Arem B (2010) The impacts of a congestion assistant on traffic flow efficiency and safety in congested traffic caused by a lane drop. J Intell Transp Syst 14 (4):197–208
5. van Arem B, van Driel CJG, Visser R (2006) The impact of cooperative adaptive cruise control on traffic-flow characteristics. IEEE Trans Intell Transp Syst 7(4):429–436
6. Hoogendoorn RG, van Arem B, Hoogendoorn SP (2014) Automated driving, traffic flow efficient and human factors: literature review. J Transp Res Board (2422), Washington, D.C., pp 113–120
7. Xiao X, Wang M, van Arem B (2015) String strategies of cooperative adaptive cruise control (CACC) vehicles for high-performance streams. Poster presented at automated vehicle symposium 2015. Ann Arbor, Michigan
8. Greenshields B, Channing W, Miller H (1935) A study of traffic capacity. In: Highway Research Board Proceedings, National Research Council (USA)
9. Geroliminis N, Daganzo CF (2008) Existence of urban-scale macroscopic fundamental diagrams: Some experimental findings. Transp Res Part B: Method 42(9):759–770
10. Daganzo CF, Geroliminis N (2008) An analytical approximation for the macroscopic fundamental diagram of urban traffic. Transp Res Part B: Method 42(9):771–781
11. Geroliminis N, Sun J (2011) Properties of a well-defined macroscopic fundamental diagram for urban traffic. Transp Res Part B: Method 45(3):605–617
12. Mazloumian A, Geroliminis N, Helbing D (2010) The spatial variability of vehicle densities as determinant of urban network capacity. Philos Trans Roy Soc London A: Math Phys Eng Sci 368(1928):4627–4647
13. VISSIM 5.40–01 - User Manual. Planung Transport Verkehr AG, 2011
14. Zhou F, Li X, Ma J Parsimonious shooting heuristic for trajectory control of connected automated traffic part I: Theoretical analysis with generalized time geography. http://arxiv.org/abs/1511.04810

Beyond Single Occupancy Vehicles: Automated Transit and Shared Mobility

Rongfang (Rachel) Liu, Daniel J. Fagnant and Wei-Bin Zhang

Abstract It is commonly accepted that Automated Transit will still be as relevant as it is now, if not more so, even when fully-automated vehicles become a reality. We need to develop a consensus on how vehicle automation will transform and perhaps disrupt the traditional transit systems, what new and different types of market-driven and publicly-run frameworks will emerge, and how we should invest our limited public resources. The two day session on Automated Transit and Shared Mobility Track (ATSM) during the 2015 Automated Vehicle Symposium (AVS) explored implications for the changing roles of transit and shared mobility as vehicle automation progresses. This chapter not only documents the main ideas presented during the symposium, but also supplements certain ideas with further discussions and clarifications after the conference.

Keywords Automated transit · Automated guideway transit · Personal rapid transit · Group rapid transit · Automated personal transit · Shared mobility

---

R. (Rachel) Liu (✉)
New Jersey Institute of Technology, University Heights, Newark,
NJ 07102, USA
e-mail: rliu@njit.edu

D.J. Fagnant
University of Utah, 110 Central Campus Drive, 2135 MCE, Salt Lake City,
UT 84112, USA
e-mail: dan.fagnant@utah.edu

W.-B. Zhang
University of California at Berkeley, Global Campus, 1357 S. 46th St.,
Richmond, CA 94804, USA
e-mail: wbzhang@path.berkeley.edu; wbzhang@berkeley.edu

G. Meyer and S. Beiker (eds.), *Road Vehicle Automation 3*,
Lecture Notes in Mobility, DOI 10.1007/978-3-319-40503-2_21

1 Introduction

Rapid development in vehicle automation technologies not only brought us hope for providing mobility for all, but also made us worry about the onset of urban sprawl when more and longer trips can be made without stressed drivers. It is commonly accepted that Automated Transit will still be as relevant as it is now, if not more so, even when fully-automated vehicles become a reality. We need to develop a consensus on how will vehicle automation disrupt traditional transit systems, what new and different types of market-driven and publicly-run frameworks will emerge, and how should we invest our limited public resources?

The two day session on Automated Transit and Shared Mobility Track during the 2015 Automated Vehicles Symposium explored implications for the changing roles of transit and shared mobility as vehicle automation progresses. This chapter not only documents the main ideas presented during the symposium but also supplements certain ideas with further discussions and clarifications after the conference.

The next section of this chapter defines Automated Transit and various members or “sub-modes” of the Automated Transit family. The third section provides a brief review of Automated Transit development and lessons learnt from the process. The forth section presents the current status of various fronts in terms of public policies, transit market shares, technologies, demonstration projects, and implementation potentials. The last but not least section explores the relationship between automated transit and shared mobility.

As such, this document will serve as a formal record identifying current and planned deployments as well as their anticipated impacts. It may be used by transit agencies, shared mobility providers, planners and policy makers in order to inform future decision-making.

2 Definitions

During the transit and shared mobility breakout session at the 2015 Vehicle Automated Symposium, attendees and participants sought to understand the changing role of transit, leading to the very question of what exactly constitutes a transit service, specifically Automated Transit. Taking a step back, it is useful to first define various transit services and identify clear examples of automated transit systems.

In this particular context, Automated Transit is passenger transportation service that is available to any person who pays a prescribed fare but is not required to be operated by driver, conductor, or station attendant [1]. As shown in Fig. 1, Automated Transit is made of a family of individual automated transit modes. So far, all of the existing commercial applications or driverless transit services can be grouped under the umbrella of Automated Transit, especially Automated Guideway

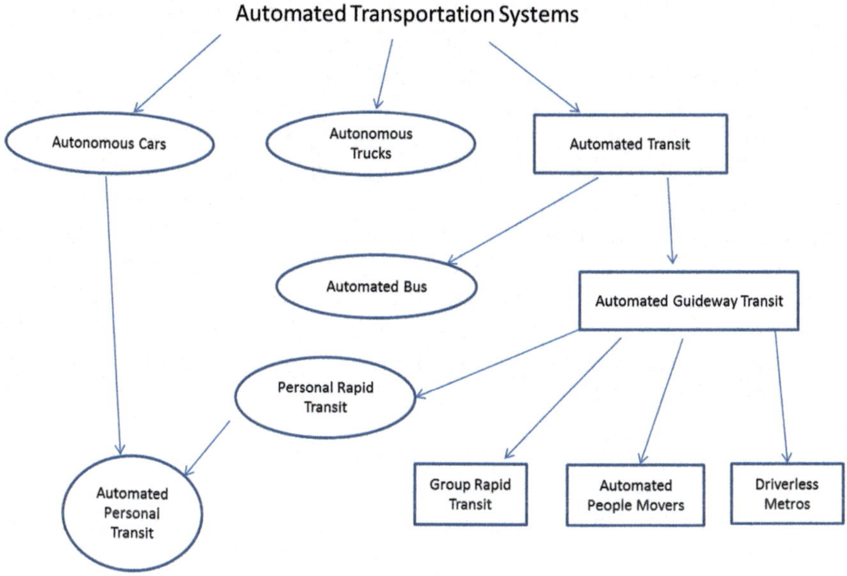

Fig. 1 Automated transit and its relationship to other modes [1]

Transit (AGT) since there is no commercial application of fully Automated Bus Transit as of 2015. However, partially automated bus transit systems with either lane guidance or speed control functions have been deployed in revenue services. These systems do require drivers to attend the system.

2.1 *Automated Guideway Transit*

Nesting under the Automated Transit umbrellas, Automated Guideway Transit (AGT) is defined as a class of transportation modes in which fully automated vehicles operate along dedicated guideways [2]. The capacities of the AGT vehicles range from 3 or 4 up to 100 passengers. Vehicles are made of single-unit cars or multiple-unit trains. The operating speeds of current systems range from 10 to 55 miles per hour (mph), and headways may vary from a few seconds to a few minutes. Automated guideway transit may be made of a single trunk route, multiple branches, or interconnected networks.

Depending on the vehicle size, capacity, and other operating characteristics, AGT may be categorized into various subgroups, such as Driverless Metros (DLM), Automated People Movers (APMs), Group Rapid Transit (GRT), and Personal Rapid Transit (PRT). Different operating environments often give AGT applications generic names, such as airport circulators or downtown people movers. Diversified track configurations, propulsion powers, and other technological features impart to AGT other names, such as monorail, duo-rail, and maglev; among others.

As illustrated in Fig. 1, all of the existing applications of automated transportation services belong to Automated Transit, especially Automated Guideway Transit (AGT), denoted by rectangles. For example, the first Group Rapid Transit Application, Morgan Town People Mover, has been operating for more than four decades without casualty or any major incidents [3]. Similar to other AV technologies, denoted in oval in Fig. 1, Automated Bus technologies are emerging and there are currently a number of demonstration projects around the word but no commercial Automated Bus applications. The Automated Personal Transit (APT) mode, to be defined in the next section, can only be realized once both AV technology and shared mobility platforms mature and intersect, and is therefore denoted by a circle in Fig. 1.

2.2 *Automated Bus*

Parallel to the definition of bus [4], an Automated Bus, or Driverless Bus (DLB), is defined as an automated vehicle designed to carry more than 15 passengers and operates on either exclusive or non-exclusive roadways [1]. Automated buses combine the advantages of automation technology with the high efficiency of public transit. When reaching a high level of automation, an Automated Bus may operate on non-exclusive roadways, where pedestrian and/or automotive traffic also exists.

2.3 *Automated Personal Transit*

With rapid development in vehicle automation (VA), it is not difficult to image what a great leap or interruption it will be when fully-automated vehicles become a reality, i.e., NHTSA Automation Level 4. According to Morgan Stanley Research [5], vehicle automation may very well develop along two diverging paths. As demonstrated in Fig. 2, the current travel scenario depicted in the first quadrant has been invaded by various shared economy pioneers such as Uber and Lyft, which are depicted in the second quadrant. The third quadrant points to the direction of automated vehicles that continue on the current private ownership axis. Far in the future, there will be the convergence of vehicle automation and shared economy—termed shared autonomy.

As one of the examples of shared autonomy, the Automated Personal Transit (APT) will have great potential to form and prosper. As an integral part of modern life in most of developed countries, a private automobile may also be one of the least utilized assets while its expense is only second to housing or shelter. If a vehicle is only utilized 1 or 2 h each day, and if the cost of hired taxi can be dramatically reduced via automated vehicles, it is quite possible that individual

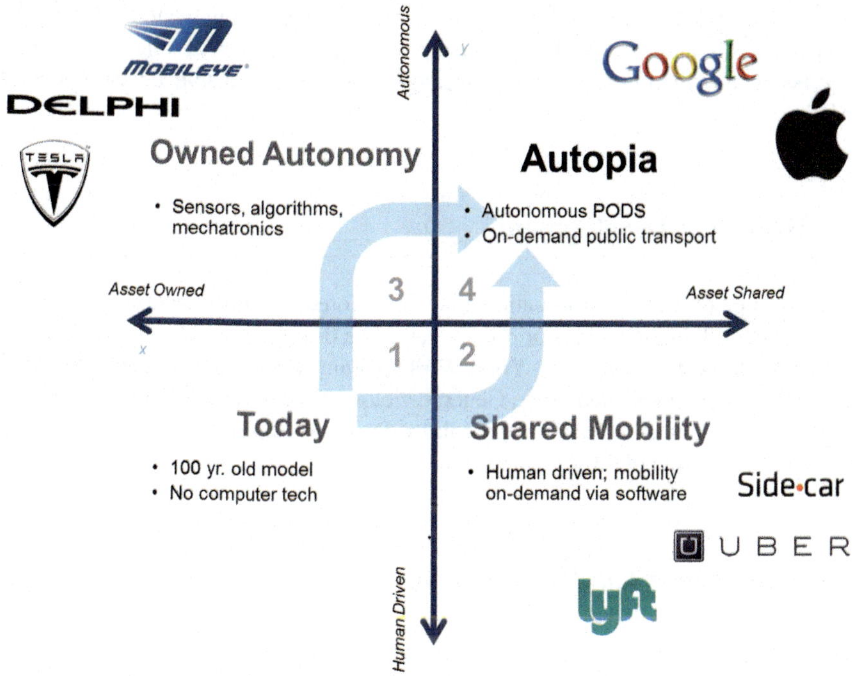

Fig. 2 Vehicle automation and share mobility path [6]

households will forego owning a vehicle altogether, while others reduce their levels of vehicle ownership. It will be much more efficient to summon an automated vehicle when one needs to travel but not have to worry about maintaining, storing, insuring, and owning the vehicle at all. This scenario will usher in a new mode, Automated Personal Transit (APT), which combines the advantages of both automated vehicles and Personal Rapid Transit (PRT), as depicted in Fig. 1.

The fleet of APT vehicles will be owned, maintained, and insured by a public or third party entity, thus transit mode. It will provide personalized, direct door-to-door service with comfort, convenience, and privacy of an automobile, thus personal, though depending on implementation, rides may be shared. An APT vehicle will be liberated from the confined tracks of PRT, the expenses of owning a private vehicle, and associated costs like parking. Instead, an APT service will possess some of the characteristics of public transit, accessible to anyone who is willing to pay a fare, and operated by a public agency over a regional network. It will also take full advantages of vehicle automation capabilities, direct door-to-door services, and reduced costs comparative to taxi, since no human driver is needed [7]. The automated or driverless features will keep the cost down and make it affordable for most travelers to hire an automated taxi—another name for APT, along with other terms like shared autonomous vehicles (SAV), autonomous taxis (aTaxis), and automated mobility on demand systems (AMODS), though these three terms are not necessarily reliant on the 'public' component of APT. The transit classification or public ownership will ensure potential funding sources, regulatory jurisdiction, and safety oversight for the sustainable development of APT.

3 Historical Development

The concept of vehicle automation may be colored by many fictions or dreamy depictions of flying machines or robotic cars. The origin of vehicle automation may be traced back to 1939 New York World's Fair where an automated highway concept was first presented at the General Motors Pavilion of the Futurama exhibit. Meantime, Automated Transit, particularly Personal Rapid Transit has been systematically documented by Fichter [8].

3.1 Automated Guideway Transit

The first AGT application was born in the US since quite a few "catalysts" worked perfectly during 1960 through 1980s [1]. First, the concept of automatic control, essential to Automated Transit, had been firmly established by the early 1950s. Second, the completion of the Apollo Moon Landing Program had freed up government funds and research capabilities, PRT had the potential and promise to fill up the plate. Third, with the fast invasion of automobiles and disappearing streetcar services, some Americans just started to question the validity of automobiles and their far reaching impact on lifestyle, environment, and society beyond.

Inspired by the Apollo Moon Landing spirit, the newly established Urban Mass Transportation Administration (UMTA), the predecessor of the Federal Transit Administration (FTA) today, made great leaps into automated transit development. UMTA not only organized technology exposition, such as TRANSPO 72 but also funded pilot projects, such as Morgan Town People Mover in West Virginia and three Downtown People Movers (DPM) in Miami, FL, Detroit, MI; and Jacksonville, FL.

Looking back, few would regard the UMTA's DPM program as a "success." Among all the three cities that implemented DPMs, Miami was often criticized for its higher initial unit costs. However, a recent examination [9] indicated that its ridership and costs closely match the original forecast, especially after the network was expanded to connect with other transit networks as originally planned, but implemented at a later stage. The DPM Program in the US was only a brief chapter as there was no more DPM application except those three pilot projects.

As the demonstration projects in the US since the 1970s faced their continuous criticism due to high cost, low ridership, and most importantly unmet expectations, AGT applications in various airports, major activity centers (MAC), and private institutions, such as amusement parks, hospitals, and museums, have been gaining steam quietly and successfully. As of 2014, there are nearly 60 Airport APM (AAPM) applications around the world [10].

While DPM and PRT applications have been riding the roller coaster of novelty thrills, government support, and disappointing implementations in the United States, AGT applications have quietly gained momentum overseas. The initial concept of a fully automated, integrated transit system in Lille, France was conceived in 1971, almost at the same time that the UMTA initiated its DPM Program.

If the very early implementation of Automated Transit technology in VAL is considered a lonely experiment with vague technology, the continued implementation of Driverless Metros in various French cities such as Lyon (1991), Toulouse (1993) and Paris (1998) has certainly solidified the pioneer position of France in embracing transit innovation and technology, and as they continually demonstrate a propensity to adapt the most advanced technologies into practical solutions. If there is any doubt about the potential of automated transit and its application in a truly dense urban area or high frequency operation systems with legacy systems, the conversion of Paris Metro No. 1 Line, the oldest and second busiest metro line in Paris, from manual operation to driverless in 2011, should have vaporizes all those doubts.

Personal Rapid Transit (PRT) was the prototype conceived by the early pioneers of Automated Transit development since the 1950s. Fitchter conceptualized the small vehicle, "Veyar", and its extensive network in an urban environment in the 1960s. UMTA attempted the PRT concept in the 1970s in Morgantown, WV, but ended up with a Group Rapid Transit (GRT) application since it utilizes much larger vehicles, a simpler network, and rarely executed direct origin to destination operations [11]. Despite many criticism and negative publicity, the hybrid Morgantown GRT has been chugging along during the past four decades and more.

3.2 Automated Bus Transit

The development of road vehicle automation began in the 1950s, when General Motors and RCA conducted experiments on automated vehicle technologies. As far as the Automated Bus Transit application, the first electronically guided transit bus was put in operation near Stuttgart, Germany, following an intensive research on vehicle lateral control by the Regional Research Lab in Germany in the 1960s and 1970s.

Under a large research program, Prédit, the French Department Transportation (DOT) committed to investigate innovative ideas for improvements of land transportation systems in the 1990s. Under this program, an optical computer vision technology was developed by MATRA, since acquired by Siemens, for bus guidance. A bus precision docking system based on optical guidance technology has been put in operation in the French city Rouen since 2001. Later the technology was deployed in Clermont-Ferrand in France. Significant saving in dwelling time was observed. Surveys of passengers have shown comparable levels of satisfaction among passengers on the electrically guided bus lines and on the LRT system.

The city of Eindhoven in the Netherlands has developed an Advanced Bus Rapid Transit System, Phileas, in collaboration with Advanced Public Transport Systems and then Frog Navigation Systems. The Phileas bus was designed ground up with integrated fully automated vehicle control capabilities. The electronic guidance system uses magnetic markers embedded in the roadway. The system was first demonstrated in 1999. However, the combination of the vehicle control system with other innovative features makes the Phileas a very complicated system to design, develop and maintain. The developers have been struggling with both the functionalities and the reliability of the automated features.

California PATH has conducted systematic research on a full array of automated road vehicle system since 1987 and demonstrated a fully automated platoon system during the National Automated Highway Systems Consortium Demonstration in San Diego in 1997. In 2003, PATH and California Department of Transportation demonstrated a fully automated bus system for a bus convoy on highway and precision docking at bus stations based on magnetic guidance technology invented by PATH.

In Japan, research and development of autonomous vehicles started in 1960s. Toyota developed an Intelligent Multimodal Transit System (IMTS) and demonstrated it in the World Expo in Nagoya in 2005. IMTS uses magnetic guidance as primary technology. Three fully automated buses were operated on a convoy on an exclusive bus way, taking passengers between the Expo centers.

While electronic guidance systems are being developed and began to see its deployment, mechanical guidance system has been deployed in several cities across the world. Essener Verkehrs, AG, the public transit agency for Essen, Germany, is the first to deploy a mechanical guidance system for transit buses on dedicated busways and the system has been in operation since 1980. Subsequently, Mannheim in Germany, West Sussex in England, and Adelaide in Australia have also followed suit. Comparing with electronic guidance, mechanical guided buses require extensive infrastructure support.

3.3 Automated Personalized Transit

Parallel efforts have been devoted to low speed, fully automated shuttle systems. A Cybercar concept was introduced by French INRIA in the 1990s. Cybercars are fully automated road vehicles that can be operated individually or in group to serve public transportation purposes. The cybercars can provide either a direct connection or operate over an elaborate network, providing on-demand door-to-door transportation. The fleet of cybercars is under control of a central management system in order to distribute transportation requests efficiently and coordinate traffic in specific settings and environment [12]. Cybercars have been demonstrated in multiple cities in Europe.

4 Current Status

In line with the theme of the Automated Vehicle Symposium, the ATSM Track has touched upon all issues related to Automated Transit, from public policy to transit share, from on-going demonstration to expansive shared economy, from to Automated Transit operation to its impact on land use and urban development.

4.1 Public Policies and Regulations

Government agencies are often expected to develop policies, create dialogue, issue guidance on standards and equity, encourage collaboration between stakeholders and conduct research that promotes integration, customer safety, reliability, and equity [13]. Policies and practices for AGT have been well developed. However, in the case of road vehicle automation or particularly in the area of Automated Transit, most government agencies are in the modes of catching up or reacting to the technology or private sector development. For example, the Mobility on Demand (MOD) Program, led by the ITS Joint Program Office (JPO) in collaboration with Federal Transit Administration (FTA), put great emphasis on connectivity and replaced "Transit" with vague "New Multimodal Mobility Concept". The US Department of Transportation (USDOT) prioritized the connected world in the order of light vehicles, then trucks, then transit. The newly developed strategic plan by USDOT [14] did not include Automated Transit in either their discussion of transit or vehicle automation.

On the regulation side, government agencies have begun to develop new regulations for autonomous vehicles. Several states in the U.S. have already published regulations for autonomous vehicles to be operated on the road in traffic for testing purposes. These regulations are applicable for Automated Transit in mixed traffic condition. However, it is yet not clear whether and what regulation will be required for exclusive operation of Automated Transit on exclusive right of way for transit corridors.

According to many transportation professionals [13, 15, 16] transit should not be the last. With exclusive right of way (ROW), 100 % market penetration, and long existing automated operations, transit has been and should have the potential to lead the pack in the path of vehicle automation. It seems that private sector is driving the technology development in this round of vehicle automation but it is strongly voiced that technology can only advance to a certain stage before it is hindered or slowed down by policy, insurance, legal, social and other related issues. Government agencies are not expected to hand out large sum of money for demonstration projects as in the past but its role of developing policy and facilitate dialogs are critically needed, especially for transit agencies and Automated Transit development.

4.2 Market Share for Transit

In the era of Connected Vehicles and Automated Vehicles (CV/AV), autonomous vehicles and Transportation Network Companies (TNC) are likely to provide travel options both complementing and competing with transit [16]. Equipped with mobile devices and widely exposed to social media, new generations of travelers are less attached to their Single Occupancy Vehicles (SOVs) and pulled more toward public transit. However; if conventional transit cannot provide safe, affordable and convenient door to door travel, many of the millennials will lose their faith in transit and turn to other alternatives.

Transit will change passively or proactively. Automation and connectivity will make transit more efficient and the newly developed multimodal transportation system will incorporate more flexible forms of transportation. For example, when automation becomes a reality, paratransit should cost much less to operate and overall mode share for transit has the potential to realize significant increases.

Discussion revealed that the transit lexicon may be further expanded when APT or privately operated transit services emerge from the combined PRT and AV technology as described in Sect. 2.3. The key is not to mince words but to understand the service concept, engineer design, and market acceptance of various shapes and forms of transportation alternatives in the new era.

4.3 Vehicle Assist and Automation

The Vehicle Assist and Automation (VAA) Program, funded by the USDOT, is one of the early exploration and demonstration projects for Automated Bus Rapid Transit technologies. The essential technology for the VAA program is based on the magnetic guidance developed in late 1980s and was demonstrated in the National Automated Highway Systems Consortium Demonstration in San Diego in 1997 [17].

The VAA initiative demonstrates Level 2 automation of steering on transit buses. The driver controls the throttle and speed and braking, while the automated system provides lane keeping and precision docking. The VAA system was tested in Eugene, Oregon. The VAA system was installed in the maintenance yard and a three mile long route.

The VAA technology uses permanent magnetic markers placed in the pavement at 1 m spacing. The costs for installing magnets are approximately \$20 to \$30 K per mile. The roadway magnetic reference system, sealed in the pavement by epoxy, is low maintenance, as the magnets are passive, reliable and more mature than other automated approaches. One bus was equipped and operated 6–8 h per day over six months in 2014. Full evaluation of the demonstration was conducted, and a report is anticipated.

Preliminary feedback from operators is positive. The VAA field operational test is the real-world deployment of automated bus in revenue service in the United States. Benefits of the VAA application include reduced stress on the driver and ability to use the bus in narrower rights-of-way.

Under the European Union's Seventh Framework Programme for research and technological development, CityMobil2 is a demonstration program of a pilot platform for automated road transport systems [15]. Citymobil2 follows the initial Cybercars concept, enabling the automated vehicles operating without a driver in collective mode. Six months demonstrations were conducted in several European cities in 2015 and 2016.

4.4 Gateway Project in UK

"Innovate UK" was launched in February 2014 to spearhead the Gateway Project in Greenwich, UK, along with two other projects: "UK Autodrive" led by ARUP and "Venturer" led by Atkin. About £19.2 Million government funding was matched by 12 consortium members to create "Innovate UK" and support three year projects. "Venturer" in Bristol is a mixture of physical and virtual environments to test sensor equipment and communication for automated vehicles, with bus comprising the primary vehicle type for data collection. "UK Autodrive" demonstrates interoperability and scalability. It uses LIDAR on carpods and LUZ Pathfinder. It also explores insurance/liability issues and identifies new business models. Finally, Gateway, led by TRL, is the consortium of members from energy, university, insurance, and car makers.

The Gateway project uses a Meridian Shuttle, which is a car-pod with an 8–10 passenger capacity. Trial 1 is the shuttle transport service in 2.2 mile route in the Meridian passing residential/commercial areas. There are shuttles serving a route from the National Maritime Museum to the Royal Observatory. Trial 2 is on the autonomous valet parking in Greenwich. In this system, participants drive to a drop-off point, get out of the vehicle, then send the vehicle off to park using a smartphone interface. Trial 3 is concentrated on urban deliveries using automated van. This trial uses Digicar to test behavior with automation and teleoperation to remotely control vehicles.

4.5 Shared Mobility by Zipcar

While the demonstration projects in prior sections represent the progress along the automation axis, the following two, Zipcar and the SMART project, showcase the development along the shared and sustainable mobility in our modern lives.

As stated by Holmes [6], Zipcar's mission is to "enable simple and responsible urban living", which has been guiding and driving the enterprise for the past

15 years. The Zipcar model is to give people the ability to live in urban areas and access to cars while freeing them from car ownership. Building on the emerging concept of shared economy, Zipcar leverages and utilizes the automobile asset across a large membership body. Since most private automobiles sit idle on average 23 h each day, there are great potentials for those under- utilized capacities to be included in the mobility spectrum.

Started with one green buggy in Cambridge Square in Boston MA, Zipcar grew into 10,000 plus cars, 900,000 members across 470 cities and towns. There are also more than 400 university campus and 50 airport Zipcar locations in the US. Collaborating with many vehicle manufacturers, Zipcar offers a wide range of fleet, about 50 makers and models. Labeled as a millennium brand, Zipcar not only saves money but also has the potential to affect auto ownership and travel behavior in the long run. Zipcar is currently a round-trip service with designated vehicle parking space homes, but is piloting one-way trips in Boston.

So far, Zipcar has been most successful in densely populated areas. The ability to support about 50 members within walking distance is the sweet spot for Zipcar selection; other transportation modes are needed for Zipcar to be effective, and propensity of population to be open to new transportation solutions are essential for Zipcar to survive.

Zipcar does not offer mobility services in isolation. It often works with transit agencies as strategic growth partners to supplement or coordinate intermodal travel for various users. Figure 3 illustrates Zipcar locations along the Redline in Boston subway systems.

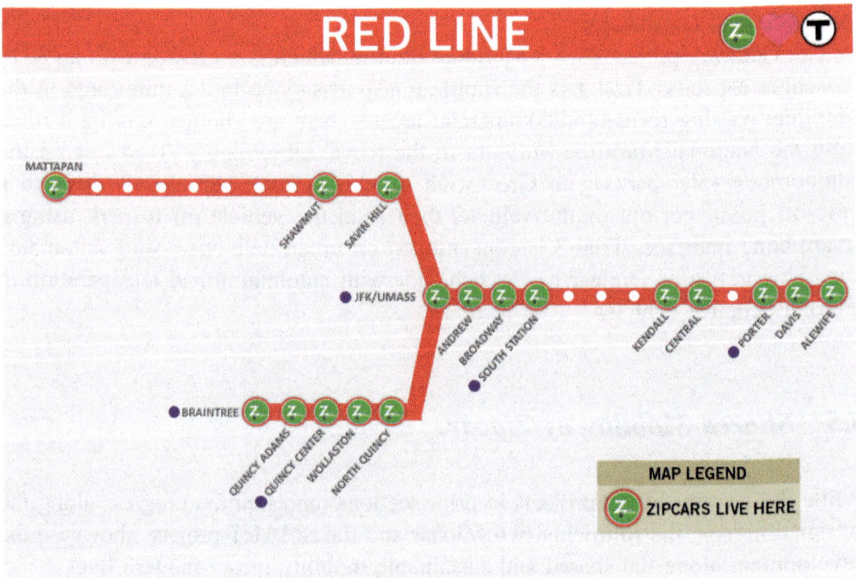

Fig. 3 Zipcar and public transit [6]

According to Shaheen [18], Zipcar users have the ability to reduce their transportation cost from 19 % of the household budget to just 6 %. As low car diet members often make more conscious decisions on travel, Zipcar users often decrease their total Vehicle Miles Travelled (VMT) in an extended time period, such as a year or a quarter, which subsequently reduces energy consumption, decreases emission and increases sustainability.

In the current business model, Zipcars are located near where members live. With Vehicle Automation, Zipcars will come to members, though their efficiency will still rely on population density, to avoid excess unoccupied travel and associated costs. Having accumulated experiences with fleet management, OEM technologies, and user interface, Zipcar will likely be partnering with more stakeholders in the new shared mobility society where community transportation solutions are woven together.

4.6 Sustainable Mobility

Expanding the shared economy and shared mobility themes, the SMART, Sustainability Mobility and Accessibility Research and Transformation, Program [19] focuses on the importance of the users and the seamless utilization of the system by people. In the fast moving, fast changing, and urbanizing transportation spaces today, there are already large quantities of infrastructure and services in place. However, those individual modes and/or elements may not be connected with each other or in the ways that serve users and/or traveler better. SMART provides a platform for initiating ideas, exchanging information, and piloting various projects related to shared mobility, connectivity and automation.

Collaborating with a large number of industries and enterprises, SMART works hard to advance connected multi-modal, IT-enabled transport systems in various locations around the globe. For example, the Veolia TRANSPORT program allows users request the super shuttle via their cell phone apps to arrange door to door transportation services. Another platform, Mobi, is a global B2B databank and network for new mobility enterprises and startups.

With great exposure to multi-culture, diverse economy and multi-modal transportation systems, the SMART experience not only opens our eyes to many, many solutions to various challenges but also made us thinking and trying to answer more specific questions:

- In what contexts do shared use, connected and automated systems make sense?
- How will shared use, connected, and automated systems be integrated within whole systems deployments both in the US and globally?
- What physical and infrastructure foundations and innovations are needed to support shared, connected, and automated systems within whole systems deployments door to door?

- What policy enablers and barriers will come into play in shared, connected, and automated systems? And related, what financing and revenue factors will come into play
- What social, psychological, and marketing factors, challenges, and opportunities will arise?

5 Relationship Between Automated Transit and Shared Mobility

The two day transit and shared mobility session of the 2015 AVS reached its climax when Alain Kornhauser, Princeton, faced off Peter Muller, PRT Consulting, in a debate moderated by Stan Young, University of Maryland. The topic centered around public investment in emerging transportation technologies, specifically a hypothetical matchup between a privately operated, fully-automated vehicle fleet, aTaxis, advocated by Alain Kornhauser, and a publically run PRT, promoted by Peter Muller.

The scenario introduced by Young hypothesized a City Council that is setting aside $1 million for the expressed purpose of a grant to facilitate testing and operation of the aTaxi fleet, or alternatively for planning and preliminary design for a PRT system. The audience of around 60 attendees affiliated with a variety of government, academia, private-sector and research institutions were invited to represent the city council to make recommendations.

Muller advanced his case, articulating arguments in support of PRT largely by contrasting with aTaxis. His arguments centered on the technology and its implementation: readiness, safety, and sustainability. Muller began by asserting the proven reliability of PRT systems, the first of which, Morgantown GRT, has been in operation since the 1970s. Muller also attacked aTaxis as an unproven product still in the testing phase, and not yet ready for implementation on any sort of scale that would provide meaningful benefits to the public at large. He next noted a zero fatality, near-perfect safety track record of PRT systems, again in contrast with the virtually unknown level of safety that may be achieved using aTaxis. He claimed that aTaxis must necessarily be less safe than PRT, since PRT removes conflicts with other vehicles and pedestrians through grade separation, while such conflicts remain when using an aTaxi fleet. Finally, Muller argued that the environmental sustainability of a PRT system was superior to that of an aTaxi fleet, asserting that PRT systems' operation on fixed guideways should provide a more efficient mobility per unit of energy, aTaxis would just add to existing traffic congestion problems, and rides in PRT vehicles would be more commonly shared.

Kornhauser's response began by questioning the viability PRT systems altogether in face of aTaxi competition. Kornhauser used the long-time existence of PRT as an indictment of the technology, citing the construction of around just one system per decade worldwide since its initial inception. On a cost-per-trip basis, he argued, PRT systems would be much more expensive than aTaxis, with significant infrastructure investment requirements; whereas an aTaxi fleet could potentially simply use the existing roads. Moreover, the superior aTaxi flexibility, i.e. an ability to travel on just about any roadway, rather than being confined to a fixed guideway, would provide a greater utility to travelers, thus boosting market share beyond what might be realized by PRT. Kornhauser asserted that aTaxis would be safer than conventional vehicles since they would effectively eliminate human error, and that while they may be involved in a crash at some time in the future, he argued that it's better to provide many people with a substantial safety improvement, rather than a dramatic safety improvement for just a few. Finally, he rebutted environmental claims by arguing that superior market share, along with shared rides, would lead to significant environmental benefits for aTaxis, while noting that PRT also carried added environmental costs in terms of new infrastructure construction.

After a brief discussion by the audience at large, Young called for a vote on the two propositions. As a result, both positions garnered above 40 %, with the privately run shared fully automated vehicle fleet winning out by a handful of votes. Yet with a number of abstentions, neither vision received a clear majority from the room, as both remained below 50 %.

6 Summary

So what exactly is public transit, in light of recent vehicle automation and other technological developments? At the 2015 AVS, many heated and passionate discussions ensued throughout the duration of the transit and shared mobility breakout session, with no firm conclusions drawn. This chapter too arrives at no definitive conclusion to this question, but rather investigates the various characteristics that help define such a system. There may be no singular criterion for what defines public transit, even though this point too was hotly debated, but rather a broad idea that public transit is a transport system that draws from among pool of key elements.

Most people's immediate conception of public transit is likely bus, metro, or light rail, systems commonly seen in daily life. Each of these examples are complete transport systems that are typically operated by a public agency, serve the general public at large, stop at pre-determined stations, and can carry large numbers of persons.

Yet recent innovations and technological developments are changing the face of transit, bringing to light this very question of what constitutes public transit. Must transit services be managed by public agencies? Must they serve the general public, or could use be restricted via membership? Does transit necessarily have to be associated with stations or physical space, or could a transit system exist as a door-to-door service? Do individual vehicles need to carry multiple unrelated travel parties? And if a given transport service connects to a larger transit system, can violations of the former criteria such as public management, unrestricted membership, station-based service, and multi-party vehicle occupancy, be acceptable? If the answer to each of these questions is no, the resulting transport mode is a private car service, an assuredly non-transit mode. Yet if the answer is no to just some of these questions, could a system still be considered transit, and if so when?

To highlight this conundrum, consider four transport systems that are generally accepted as various forms of transit: automated transit networks (ATNs) or PRT, paratransit, privately-run bus, and demand-responsive feeders for line haul mass transit. An ATN may be implemented using a single-party occupancy framework, thus violating the above noted occupancy condition. Paratransit usage is typically restricted to qualified riders, and operates using a door-to-door framework. A privately-run bus system is by definition not run by a public agency. The demand-responsive feeder may be set up to connect single parties to the line haul system, and may use door-to-door service on the non-station end. But if the feeder becomes run by a private entity does it cease to become a form of public transit? What if it restricts use to pre-qualified members, similar to how car-sharing companies like Zipcar and Car2Go operate?

Moreover, as automated on-demand shared use vehicle fleets look set to become a reality, the distinctions between shared mobility and transit systems look to become further blurred. Transportation network companies, such as Uber and Lyft, and carsharing companies, such as Zipcar and Car2Go, may be considered examples of shared mobility systems. Yet suppose their fleets eventually become fully automated and suppose a transit agency similarly determines that it can more effectively serve the public through single-party fully automated vehicles, rather than through a conventional or even automated bus service. While organizational objectives may differ, the only functional distinction may be the public versus private ownership. If that is the case, is this enough of a distinction to declare one form transit and the other not?

In closing, it should be noted that these distinctions between what does and does not constitute transit have real world impacts. The regulatory and public funding environment for transit is dramatically different from the shared mobility space, yet clarity is becoming increasingly difficult due to rapid developments in mobility platforms and may become even more difficult to discern as the pace of vehicle automation accelerates. However, regardless of what new types emerging transportation modes are classified transit, the one near-certainty is that they will bring a host of new alternatives and opportunities to the traveling public.

References

1. Liu R (2016) Automated transit: planning, operations and applications. Wiley
2. Liu R (2010) Spectrum of automated guideway transit (AGT) technology and its applications. In: Kutz M (ed) Handbook of transportation engineering. McGraw-Hill
3. Raney S, Young S (2005) Morgantown people mover—updated description. In: Proceedings of transportation research board annual conference, 2005

4. United States Government Publishing Office, Code of Federal Regulations. http://www.gpo.gov/fdsys/pkg/CFR-2011-title49-vol5/xml/CFR-2011-title49-vol5-part393.xml, 2011
5. Morgan Stanley Research (2015) Autonomous cars: the future is now. January 23, 2015. Accessed October 2015
6. Holmes J (2015) Zipcar—The promise of automated vehicles. Presented at the 4th automated vehicle symposium, Ann Arbor MI, July 21, 2015
7. Fagnant D, Kockelman K Dynamic ride-sharing and optimal fleet sizing for a system of shared autonomous vehicles in Austin, Texas, Forthcoming in Transportation
8. Fichter D Individualized automated transit and the city. Chicago, B. H. Sikes
9. Cottrell, W (1955) Moving driverless transit into the mainstream - research issues and challenges. Transportation Research Record: J Transport Res Board 1(1)
10. Fabian L (2014) Airport APMs, Trans. 21 Newsletter, Fall 2014
11. National Research Council, Summary of a Review of the Department of Transportation's Automated Guideway Transit Program. Contract No. OT-OS-40022, 1975
12. Parent M, Arnaud de La F (2005) Cybercars – past, present and future of the technology. In: Proceedings of ITS World Congress, 2005
13. Lesh M (2015) Multimodal ITS transit research direction – integrated, connected, on-demand shared use mobility". Presented at the 4th automated vehicle symposium, Ann Arbor MI, July 21, 2015
14. US Department of Transportation, Beyond Traffic 2045, Trends and Choices. 2015
15. Alessandrini A, CityMobil2 – initial results on automation for shared mobility. Presented at the 4th automated vehicle symposium, Ann Arbor MI, July 21, 2015
16. Polzin S (2015) How new technologies and autonomous vehicles may change public transportation. Presented at the 4th automated vehicle symposium, Ann Arbor MI, July 21, 2015
17. Rajamani H, Tan S, Law B, Zhang WB (2000) Demonstration of integrated longitudinal and lateral control for the operation of automated vehicles in platoons. IEEE Trans Control Syst Technol 8(4)
18. Sheehan S (2015) Transportation sustainability, disrupting mobility summit. Boston
19. Zielinski S (2015) Whole system connectivity: focus on user. Presented at the 4th automated vehicle symposium, Ann Arbor MI, July 21, 2015

Vulnerable Road Users: How Can Automated Vehicle Systems Help to Keep Them Safe and Mobile?

Alma Siulagi, Jonathan F. Antin, Lisa J. Molnar, Sue Bai, Seleta Reynolds, Oliver Carsten and Ryan Greene-Roesel

Abstract For the first time in 2015, the Automated Vehicle Symposium featured a breakout session explicitly devoted to vulnerable road users (VRUs) and their use of and interactions with automated vehicles. A number of stakeholders, experts, and researchers from a variety of fields presented and discussed the state of current research and thought concerning the potential relationship of vulnerable road users

A. Siulagi
Design City and Regional Planning, University of Pennsylvania,
4635 Sansom St. #2, Philadelphia, USA
e-mail: siulagia@design.upenn.edu

J.F. Antin (✉)
Virginia Tech Transportation Institute, 3500 Transportation Research Drive,
Blacksburg, VA 24060-0536, USA
e-mail: jantin@vtti.vt.edu

L.J. Molnar
University of Michigan Transportation Research Institute, 2901 Baxter Road,
Ann Arbor, MI 48109-2150, USA
e-mail: ljmolnar@umich.edu

S. Bai
Honda R & D Americas, Inc., 1000 Town Center Dr., Ste 2400,
Southfield, MI 48075, USA
e-mail: SBai@oh.hra.com

S. Reynolds
Los Angeles Department of Transportation, 100 S Main Street, 10th Floor,
Los Angeles, CA 90012, USA
e-mail: seleta.reynolds@lacity.org

O. Carsten
Institute for Transport Studies, Leeds LS2 9JT, UK
e-mail: o.m.j.carsten@its.leeds.ac.uk

R. Greene-Roesel
San Francisco County Transportation Authority, 1455 Market Street, 22nd Floor, San Francisco 94103, USA
e-mail: ryan@sfcta.org

G. Meyer and S. Beiker (eds.), *Road Vehicle Automation 3*,
Lecture Notes in Mobility, DOI 10.1007/978-3-319-40503-2_22

and automated vehicles and how to maximize the benefits this novel technology might bring to these individuals. The topics included the role of design, various technological solutions, policies, and programs that could advance the safe mobility of VRUs in a future with an integrated fleet of automated vehicle systems. Through expert-led small group discussion, the breakout group produced a list of possible definitions for VRUs including pedestrians, cyclists, seniors (pedestrians as well as drivers), and identified key research gaps within the context of this multifaceted segment of the population. Some of these gaps related to motorcycle interactions, how different groups of VRUs will accept emerging AV technologies, and goals and solutions when considering how best to share limited roadway space across all road user constituencies.

Keywords Aging drivers · Pedestrians · Bicyclists · Automated/autonomous surface vehicles · Automated vehicle symposium · Vulnerable road users · Connected technologies · Automated rules following · Unintended consequences · Privacy concerns · Traffic flow · Holistic inclusive platform · Vision zero · Road sharing

1 Introduction

As automated features begin to permeate the market and the possibility of partial and one-day full automation becomes more likely, it is essential to both review possible impacts on traditionally vulnerable populations as well as reimagine vulnerability under this emerging market.

The breakout session, "Vulnerable Road Users: How Can Automated Vehicle Systems Keep them Safe and Mobile?" held at the Automated Vehicles Symposium 2015 in Ann Arbor, Michigan attempted to explicitly address these pressing questions for the first time at the AV symposium series. The 3-h session aimed to define vulnerable road users (VRUs) within the context of automated vehicles (AVs); to identify critical research gaps; and to determine whether the topic of VRUs was suited for future Automated Vehicle Symposium breakout sessions. Experts from across sectors and disciplines took turns presenting. Then, an open-ended discussion was held with the moderator, speakers, and audience members participating vigorously. Together, the session's experts and participants produced a list of possible definitions of VRUs, a list of research gaps, and agreed on the importance of continuing the conversation in the coming years.

Following is a synthesis of the perspectives of the expert presentations, a brief analysis of the benefits and challenges identified in the presentations, a review of the insights from the panel and small group discussion, and the results of the session, including a list of possible definitions and critical research gaps. The information, analysis, products, and conclusions from this breakout session provide a much-needed baseline for future discussions centered on AVs or VRUs.

The term "vulnerable road users" has traditionally been limited to pedestrians, bicyclist, and motor cyclists because these groups are considered to be at the highest risk in traffic due to their lack of external protection [1]. However, in recent years, it has been argued that older drivers also fall into this category because of their increased fragility and frailty which make them more susceptible to injury and death in the event of a crash [2]. Efforts to reduce injuries and deaths among vulnerable road users have been multifaceted and include improvements in roadway infrastructure, education and training relative to road user behavior, implementation of policies and regulations such as helmet laws, and advances in vehicle design [3, 4]. Recent advances in automated vehicle technology also show promise for enhancing the safety of all road users, including vulnerable road users. However, these technologies are still early in their development and important questions remain about their effectiveness in enhancing the safety and mobility of vulnerable road users, and how they might be used and accepted by these populations.

2 Issues and Solutions

The presentations in the breakout session represented a wide variety of experts and perspectives of those working on projects reaching across sectors, disciplines, and the globe. The diversity of perspectives also meant a wide range of definitions of who qualified as a VRU—the presentations mentioned bicyclists, pedestrians, older drivers, and urban populations as possibly vulnerable within an AV context. Each presentation touched upon current projects, possible benefits, challenges, and future steps forward.

2.1 *The Presentation Summaries*

2.1.1 Vehicle-Smartphone Communication to Protect Road Users (V2P and Beyond) Presented[1]

Ms. Bai presented current research scenarios being conducted by Honda and Qualcomm that explore how connected vehicle technology could extend to pedestrians, bicyclists, and other VRUs via smartphone technology. This would address pedestrian (and other VRU) deaths and their societal cost. However, it will require cooperation between automotive manufacturers and the wireless industry to develop the standards necessary for wider production. Ms. Bai stated that this technology will be tested as a pilot program on a campus in the near future. Possible

[1]Presented by Sue Bai, Principal Engineer, Automobile and Technology Research Department, Honda R&D America, Inc.

limitations included identifying relevant pedestrians to vehicles, sudden crossing maneuvers, positional accuracy, positional data in parking structures, smartphone battery power, and communication channel congestion.

2.1.2 How Can We Use AVs to Protect VRUs?[2]

Dr. Van Houten presented common risk and crash scenarios for pedestrians and bicyclists that AV systems would be able to address by simply automatically following the rules currently in place (which human drivers may or may not follow). He offered the example where at a stoplight a human driver might pull forward past the stop bar, whereas an AV could be programmed to stop appropriately behind the line, thus keeping crossing VRUs safer. Dr. Van Houten also mentioned personal GPS transponders as a possible way for bicycles, pedestrians, and cars to communicate to prevent accidents. However, he expressed concerns that these solutions also had possible negative effects which would need to be addressed: privacy, possible negative influences on traffic flow, and dependence on smartphone technology (as per Ms. Bai's presentation) among them.

2.1.3 Automated Vehicle Symposium Vulnerable Users[3]

Ms. Reynolds presented a way for cities to approach the oncoming changes in transportation, with AVs as one of many possibly "disruptive" forces. She outlined a role for the city in promoting the outcomes best for their urban populations at large, in particular safety and efficient mobility. Ms. Reynolds asserted that cities are still learning to be start-ups and can leverage their strength in street design and managing mobility by integrating modes into a holistic and inclusive user platform. Cities may not provide the technology platforms for new mobility, but they should lead the policy and regulatory agenda with these top level outcomes in mind. She stressed the need for city representatives to be engaged early in such conversations regarding possible urban transportation futures. In terms of AVs, Ms. Reynolds mentioned two possible outcomes: the "robo-taxi pod car" and the "luxury long commute" model of AV deployment, asking cities and their leaders to investigate which model would bring cities closer to vision zero goals. She mentioned that the repurposing of roadway space can generate powerful backlash and need strong partnerships in both the public and private sphere to determine outcomes.

[2]Presented by Ron Van Houten, Ph.D., Professor of Psychology, Western Michigan University.
[3]Presented by Seleta Reynolds, General Manager, Los Angeles Department of Transportation.

2.1.4 Implications of Automated Vehicles for Older Drivers

Dr. Molnar presented her research on how AVs may affect older drivers. She noted that older drivers in the US, like other age groups, prefer to travel by personal vehicle but may experience age-related declines in visual, psychomotor, and cognitive abilities that can compromise safe driving. Thus, the emergence of AVs may be a great boon for aging drivers, allowing them to maintain mobility while reducing their risk of crashes. At the same time, Dr. Molnar noted that older drivers may face special challenges in using and benefiting from AVs. She discussed several of these challenges and highlighted the importance of making sure that AV design is responsive to the needs and preferences of older adults. She emphasized that design features such as smart headlights, reverse monitoring systems, assistive parking systems, and lane departure warnings work best for older drivers.

2.1.5 Automation Connection and Cyclists[4]

Dr. Carsten presented on his work on bicyclists as VRUs and the challenge of accommodating them in AV design. He outlined common bicycle conflicts with cars and their unique modal characteristics that make them difficult to design for, such as their speed relative to prevailing motorized traffic and tendency to violate traffic rules. Dr. Carsten showcased the XCYCLE program, which reviews how connected vehicle technology can promote cyclist safety via passive, two-way, or active detection systems. The central question remains how the connection can be imposed on cyclists.

2.2 Summarizing Proposed Solutions to Accommodate VRUs Within an AV Context

These five experts presented the state of research, programs, and pilots currently addressing the complexities associated with AVs in the context of servicing or interacting with VRUs. Table 1 organizes the solutions touched in in presentation into four general categories: AV design, additional technologies, environmental design, and programs and policies. Some solutions are combined with others. For each solution, a goal is presented, as well as the limitations and steps forward that the presentations discussed. Many of these issues are discussed more thoroughly in the section summarizing the panel and small group discussion.

---

[4]Presented by Oliver Carsten, Ph.D., Professor of Transport Safety at the Institute for Transport Studies, University of Leeds.

Table 1 Summary of the approaches to accommodate VRUs within an AV context

Proposed solutions	Presenter	Goal	Limitations	Steps forward
AV design				
Automated warning system paired with pedestrian detection system	Bai	To reduce the risk of injury or fatal pedestrian injuries	Possible disruption to traffic flow	Honda and Qualcomm campus pilot
Designing AVs to follow existing traffic rules (speed, roadway markings)	Van Houten, Carsten	To reduce the risk of injury or fatal pedestrian injuries	Public acceptance, as many drivers today consistently choose not to obey may traffic laws and regulations	
Designing an elderly friendly AV user interface for seniors AVs	Molnar	To reduce the risk of distraction for older drivers	May not be desirable if marketed as being an "older driver" vehicle or feature	Suggested further research
Additional technologies				
Pedestrian detection systems using either GPS transponders, smartphone technology, or connected technologies	Van Houten, Bai, Carsten	To reduce the risk of injury or fatal pedestrian injuries	Privacy, dependence on smartphone technology	Honda and Qualcomm campus pilot; XCYCLE program (H2020 smart green and integrated transport)
User interface integrated across all available modes	Reynolds	To increase mobility options for urban residents and move towards transportation as a utility	Not discussed	
Environmental design				
Redesign infrastructure and roadways to accommodate modes differently	Reynolds	To reduce the risk of injury or fatal pedestrian injuries on previously identified dangerous streets	Lack of public acceptance; concern about loss of private parking	

(continued)

Table 1 (continued)

Proposed solutions	Presenter	Goal	Limitations	Steps forward
Programs and policies				
Cities as key policy-makers in decisions that may disrupt the transportation system	Reynolds	To enable cities to achieve triple bottom line outcomes for urban residents around safety, mobility, sustainability, and resilience	Cities wield limited power and must leverage outcomes to achieve desired or required levels of efficiency	Cities getting involved in and establishing role in decision-making conversations concerning urban transportation systems' futures
Technology adoption program tailored to older adults	Molnar	To increase the quality of life for aging drivers by increasing their willingness to engage with novel emerging technologies	Not discussed	Suggested further research

3 Issues and Current Thought

The questions addressed in the panel and the small group discussions identify important issues with current thought surrounding AVs and VRUs. A brief summary of each is presented below to stimulate discussion and broaden the scope of the issue beyond expert presentations.

3.1 Review of the Panel Session

In the panel session, participants asked a wide range of questions that focused on proposals to alter roadway design, the Honda-Qualcomm pilot project, and ways to design programs and policies that could achieve preferred safety outcomes.

Following up on Ms. Reynolds and Dr. Van Houten's presentations, participants questioned how contemporary roadway infrastructure would interact with AVs and VRUs. Particularly, one audience member raised an issue that has been debated previously in AV circles: would AVs be programmed to recognize and obey speed limits? This would bring certain benefits to pedestrian and bicyclists and perhaps other VRUs. While Dr. Carsten took the line that lowering speed is a basic safety principle and that AVs should not allow excessive speeding, Ms. Reynolds pointed out that how we set speed limits is a historical process already divorced from both current safety realities and possible AV futures. In this case, it would seem that the contemporary traffic safety infrastructure should be reconsidered within current and possible future contexts including automation.

Another audience member followed up with a question about Ms. Reynold's proposal to redesign streets to accommodate non-motorized modes and how devoting more road space to bicycles and pedestrians might influence congestion and throughput for vehicles. Ms. Reynolds supplied Portland, Oregon and San Francisco as examples of cities that have dedicated significant lane space to bicyclists without degrading throughput. She explained that in corridors with congested vehicle traffic, cities and decision-makers should aim to draw drivers into other modes rather than jumping immediately to increasing capacity for vehicles.

A second group of participant questions focused on the Honda /Qualcomm pilot for smartphone pedestrian detection systems. One questioned how the system design would respond to irrational users, such as pedestrians who do not pay attention or cyclists who take safety risks. All of the panel members agreed that there is no known way to design for this, and that while that question has yet to be answered, the system should be forgiving of all behavior. A panel member suggested that if pedestrians could cross where they liked without the risk of injury, this would change much for populations such as the blind.

Other participants questioned whether smartphones were the right technology to support pedestrian and bicyclist detection systems. Access to smartphones, the unreliability of smartphones (e.g., battery life), and the suitability of the interface to relaying critical and immediate safety information were raised as potential system failures.

Participants also expressed concerns about inducing safety outcomes both at the urban and subpopulation level. Ms. Reynolds discussed the "vision zero" strategy adopted by five cities in California, and their different approaches to creating safer streets, including identifying the most important crash sites and streets, looking for patterns amongst crash data, and re-conceptualizing project delivery, partnerships, and the role of the citizen. Dr. Van Houten pointed out that in the case of requiring AVs to obey traffic safety laws, safety outcomes are built into AVs. Two other participants wondered how children's safety would work in an AV future, and how the benefits could be maximized for the elderly despite their tendency towards slow adoption of technology.

3.2 Questions for Further Discussion

Due to the size of the audience, small group discussion was conducted as a single group in a less formal manner than the panel-audience set up. Dr. Antin began by reading the discussion questions the breakout session:

- Who are the key vulnerable road users that could be affected by AV technology?
- What are the unique needs of each VRU group?
- How might each group benefit from AV technology?

Table 2 VRU candidates and qualifications

VRU candidates	Qualification
Pedestrians	Physical vulnerability and intersecting or shared right-of-way with vehicles
Bicyclists	Physical vulnerability and intersecting or shared right-of-way with vehicles
Motorcyclists, scooters, electric bikes	Physical vulnerability and intersecting or shared right-of-way with vehicles; however they present additional challenges such as white-lining behavior
Skateboarders, Segway riders	Physical vulnerability and intersecting or shared right-of-way with vehicles
Seniors	Physical vulnerability and prone to cause more collisions
Disabled	Depends on the particular disability (e.g., slower as a pedestrian, deafness, or blindness)

- Do AVs pose any risks to VRUs?
- What are key research gaps in exploring the relationships between VRUs and AV technology?

These questions framed the results presented below in Table 2. However, the small group discussion included ideas for the future of VRUs as a topic of focus at AVS.

3.3 The Future of VRUs at AVS

Participants reviewed whether VRUs should be included as an ongoing topic at AVS considering its new place on the program and relatively low turnout. While many felt the session was important, others suggested that it could be merged with the popular, two-day human factors session. This would mean VRUs could gain a wider audience and consider breaking out into its own session in later years.

4 Results

One of the key topics of discussion was who (or what transportation modes) qualified as VRUs? Different ideas were discussed; these are presented in Table 2.

The questions addressed in the panel and the small group discussions identify important issues with current thought surrounding AVs and VRUs. A brief summary of each is presented below to stimulate discussion and broaden the scope of the issue beyond expert presentations.

- How do motorcycles mesh and interact with automated vehicle functions?
- How do we define what constitutes a VRU and is it fundamentally one group or many with varying needs and concerns for each?
- How will the various VRU constituents accept emerging AV technologies?
- What are the goals and solutions when considering how best to share limited roadway space (e.g., goals may include safety, fairness, costs, protecting the environment, space livability, etc.)?
- How can AV technologies facilitate safe mobility for senior drivers, and what are the hurdles that must be overcome when considering this growing segment of our population within this context?

5 Discussion and Conclusions

This breakout session was noteworthy for three reasons. First, this was the first formal breakout session at the annual Automated Vehicle Symposium devoted explicitly to the needs and concerns of VRU. Second, it brought together decision-makers and researchers across a broad spectrum who may not have previously communicated or discussed common issues surrounding VRUs from their unique perspectives. And lastly, this represents the first attempt to begin an ongoing, cross-sectoral conversation that defines VRUs in an ever-evolving AV context.

References

1. Organization for Economic Co-Operation and Development, Safety of Vulnerable Road Users, DSTI/DOT/RTR/RS7(98)/FINAL, Organization for Economic Co-Operation and Development, 1998
2. McMahon K (2008) Vulnerable road user safety: children, elderly road users and pedestrians. International Transport Forum, Joint Transport Research Centre, Paris, France, 2008
3. Constant A, Lagarde E (2010) Protecting vulnerable road users from injury. PLoS Med 7(2):1–4
4. Goodwin A, Kirley B, Sandt L, Hall W, Thomas L, O'Brian N, Summerlin D (2013) Countermeasures that work – a highway safety countermeasure guide for state highway safety offices, 7th edn. University of North Carolina Highway Safety Research Center

Implications of Vehicle Automation for Planning

Sivaramakrishnan Srinivasan, Scott Smith and Dimitris Milakis

Abstract The substantial uncertainty associated with the capabilities and deployment time lines of automated vehicle (AV) technologies makes it difficult to consider AVs in the long range transportation planning process. At the same time, given current and anticipated resource constraints, the consideration of AV technology could be critical for developing efficient and sustainable transportation systems. This paper documents findings from a workshop of modelers, planners, and researchers on (1) potential uncertainties associates with AV technology and adoption, (2) its implications for the transportation planning process, and (3) possible approaches (including immediate steps) that can help address planning under uncertainty. The workshop was held in the context of the Automated Vehicle Symposium 2015.

Keywords Long range transportation planning · Uncertainties · Scenario planning · Pilot projects · Education

S. Srinivasan (✉)
University of Florida, 365 Weil Hall, Box 116580, Gainesville, USA
e-mail: siva@ce.ufl.edu

S. Smith
Volpe National Transportation Systems Center, U.S. Department of Transportation, 55 Broadway, Cambridge, MA 02142, USA
e-mail: scott.smith@dot.gov

D. Milakis
Department of Transport & Planning, Delft University of Technology, Stevinweg 1, PO Box 5048, 2628 Delft, CN, The Netherlands
e-mail: d.milakis@tudelft.nl

G. Meyer and S. Beiker (eds.), *Road Vehicle Automation 3*,
Lecture Notes in Mobility, DOI 10.1007/978-3-319-40503-2_23

1 Why Should Planners Care Now? Workshop Motivation and Organization

Metropolitan Planning Organizations (MPOs) in the Unites States routinely undertake long-range transportation planning to help shape the land use and transportation systems of the region that meet the goals of the residents. The time horizon of these plans is typically 25 years or more into the future. Over the same period, the automated vehicle technology is expected to be significantly mature and widely deployed. Yet, little attention has been given to the potentially transformative impact this technology could have on the long-range transportation planning procedures and practice. Arguably, one of the key issues leading to the lack of attention is the substantial uncertainty associated with capabilities and the deployment time lines of the AV technology. At the same time, given the current and anticipated resource (finances, land, fuel, etc.) constraints, consideration of the AV technology could be critical for developing efficient and sustainable transportation systems.

In the context of the discussion presented, the goal of the workshop was to assemble a group of modelers, planners, and researchers to discuss (1) potential uncertainties associated with AV technology and adoption, (2) its implications for the transportation planning process, and (3) possible approaches (including immediate steps) that can help address planning under uncertainty.

To achieve the above goals, the workshop was broken down into two sessions. The first session "Technology, policy and user issues" covered key uncertainties for planners with three brief presentations and audience discussion to identify the most important driving forces and potential implications of AVs for transportation demand forecasting and planning. The second part "Scenario development process and implications for planning" introduced scenario-based planning as an approach for dealing with uncertainty with a case-study presentation, host-invited response from regional and state transportation officials, and concluded with an audience discussion.

Material presented in this paper is drawn from the presentations and audience discussions at the workshop. It is important to emphasize that the goals of the workshop, and hence this paper, are to identify key questions from the standpoint of long-range transportation planning and to suggest initial steps and methods that can ultimately lead to the development of appropriate planning procedures for an automated vehicles world. The paper itself does not provide a recipe for AV demand forecasting and planning.

The rest of this paper is organized as follows. Section 2 presents a discussion of the major uncertainties associated with automation and its implication (key questions) for planning. Section 3 focuses on what planners can do now. Starting with scenario-based planning as an approach for undertaking a structured discussion about an uncertain world, the section also outlines the need for pilot studies and education as immediate needs identified by workshop attendees. Section 4 provides a short summary of the paper and identifies major conclusions from the workshop.

2 Major Uncertainties with Automation and Implications for Planning

2.1 Factors Causing Uncertainty

Three major areas of uncertainty will influence how automation affects the transportation system during the next 25 years. These are technology, policy, and user attitudes and behavior.

Four technological elements [1] are needed for wide AV deployment: telematics, ADAS (advanced driver assistance systems), ad hoc communications (DRSC, LTE-D) and self-driving. Although driver assistance systems exist today, they will not have transformative impacts on transportation demand, because a qualified driver needs to still be engaged. One can also envision high automation (SAE level 4; see [2]) operating in certain conditions and on some roads within the next 25 years. However, it is not clear that full automation (SAE level 5), will be widely available within the next 25 years.

The second factor is policy. Planning and policy are intertwined [3] with the new urban mobility (shared cars and rides) raising some significant policy issues. At this point, the United States has been blessed by having comparatively few regulations. That said, perhaps planners should be more active in addressing policy issues, in order to make desirable outcomes more likely. A few questions that planners might consider include the following: How fast does automation happen? What fraction of the fleet has to automate before you start to see benefits? How does this interact with funding? What impacts should we expect on vehicle ownership, VMT, and land use? What does this mean about what we should build in terms of infrastructure?

The third factor, user attitudes and behavior, will significantly affect how automation impacts the transportation system [4]. Key questions include (1) will users embrace automation and (2) will users be willing to share vehicles, riders and data? Further, planners must always keep in mind the "who"—we traditionally think of the user as a person or household, but the user might also be a freight provider or fleet manager. In terms of willingness to adopt automation, important factors include capital cost, per trip cost, willingness to cede control (driving or routing), perception of value (safety, convenience, multitasking). These factors will influence market penetration (percent owned or shared), changes in routing and changes in travel disutility. In terms of willingness to share vehicles, rides and data, important factors include relative cost differentials between owning and sharing, use of vehicles for transport versus storage, convenience of ride-sharing (wait time, routing), willingness to share in-vehicle space with others (safety, privacy) and data privacy concerns. Based on several surveys conducted to date, Smith [5] notes that 20–40 % of respondents indicate a willingness to purchase an autonomous vehicle while 50–70 % of respondents indicate a willingness to ride in one.

2.2 Key Questions for Travel Demand and Planning

The uncertainties associated with the AV technology and its deployment leads to several critical questions about its impact on travel demand and planning decisions. Some of these are discussed next.

Will demand decrease or increase? Several aspects of AV technology and deployment models might suggest a decrease in vehicular travel demand. For instance, fewer vehicles might be needed to move travelers, and more so if the society moves away from ownership of vehicles to subscription to car/trip sharing services. As fully automated vehicles determine optimal paths for travel, the trip lengths might decrease. Further, the millennials (who would be the majority of the traveling population 15–25 years from now) seem to have less preference for suburban living than previous generations. At the same time, AV technology opens up new mobility options to children, elderly, and disabled which could potentially increase overall travel demand. Further, since travel time in automated vehicles can be productive time, people may become less sensitive to longer trips (assuming that the cost of these trips are not substantial). Fully automated vehicles may also have to make several dead-heading (zero occupant) trips adding to the overall travel demand of the system.

Will capacity increase? AVs promise safety at higher speeds and therefore the transportation system can be expected to be faster and have fewer incidents. Further, the headways between vehicles can also be shorter leading to increased capacity. Signal timing plans can be further optimized (or maybe even eliminated) recognizing better "reaction times" of vehicles than humans. At the same time, during the transitionary period, the mixed-vehicle fleet can lead to decreased speeds and increased headways to allow for safe interaction of human-driven vehicles with self-driven vehicles. Finally, the performance of AVs in bottleneck situations (which essentially define capacity) is still unknown.

What level of automation should we be planning for (now)? While discussions about AVs often (implicitly) focus on a future of fully autonomous vehicles, it is important to realize that both NHTSA and SAE recognize increasing levels of automation. Using the SAE levels of automation, perhaps we should be looking at levels 2/3 (some controls automated, driver's role is still present) first as opposed to level 5 (fully autonomous) now?

What does automation mean for infrastructure investment? The investment decisions we are making today may last 70 years into the future. While automation may provide opportunities for savings which is important in financially-constrained times, allocating funds to deal with uncertain AVs may also be a risky proposition. It is important to consider at least for three major types of infrastructure: roadways, transit, and the ITS infrastructure. In the context of roadways, it is important to consider which facilities are most likely to see AVs and correspondingly start developing those for a future of AVs.

The future perception of transit is of interest. While shared AVs might substitute for some bus services, high-capacity fixed-guideway heavy rail systems may still be needed on some corridors. Therefore, investments in AVs may be made instead of investments in transit or to complement investments to improve a transit system. It is also important to consider the impacts of automation on the profession of commercial drivers while making decisions.

Finally, the nature of ITS infrastructure needed will also be different for AVs. Broadly, while current ITS provides information to humans (say a variable message sign), ITS for AVs should be designed to communicate directly with the vehicles and also be capable of receiving and processing information transmitted by the vehicle. Above all, AVs are likely to generate enormous amounts of data and we should have the infrastructure to be able to store and analyze these.

What about other "Mega-Trends" and Land Use Scenarios? MPOs may be experiencing their own unique socio-economic-attitudinal trends which need to be considered as a complement to AV projections. It was also observed that most US cities have lower densities of development which may not drastically change into the future. Simulation studies of AVs in low-density areas shows that the occupancy levels of these vehicles might be low, degrading their effectiveness in reducing travel demand. Thus, AVs must be visualized in the context of the local land-use trends and policies.

3 What Can Planners Do Now?

This section of the paper discusses some of the steps that planners can take now towards incorporating AVs in the long-range transportation planning process. Specifically, scenario planning is introduced as a structured way of dealing with uncertainty. As the audience noted, while uncertainty in demand forecasting is not new, the magnitude of uncertainty introduced by AVs is significantly higher and the consequences of incorrect decision making can also be substantial. Section 3.1 introduces scenario planning and presents a case-study from the Netherlands as an example of how these may be applied to study planning for AVs. Section 3.2 presents the need for pilot projects and Sect. 3.3 stresses the need for education as next steps.

3.1 Scenario Planning

3.1.1 Introduction

Scenario analysis can help planners accommodate uncertainty of the future by identifying different plausible states of the world and possible impacts on the issue under question [6]. Scenarios offer more flexible and creative outlines of the future than probabilistic forecasts. In the following section, we present a scenario study that identified plausible future development paths of automated vehicles in the

Netherlands and estimated potential implications for traffic, travel behavior and transport planning on a time horizon up to 2030 and 2050. We first describe the methodology and then briefly present the scenarios and conclusions about potential development and impacts of automated vehicles (see Milakis et al. [7] for more details about this study).

3.1.2 The Dutch Case Study

The scenario study involved experts from various planning, technology, and research organizations in the Netherlands and was completed in three workshops. The methodology comprised five sequential steps. The first step involved the identification of the key factors and driving forces of the development of automated vehicles in the Netherlands. The impact and uncertainty of those driving forces were subsequently assessed, followed by the construction of the scenario matrix. The penetration rates and potential implications of automated vehicles in each scenario were then estimated. The study was completed with an assessment of the likelihood and overall impact of each scenario.

Four scenarios were built around permutations of two driving forces of automated vehicles: technology and policies. These driving forces were assessed as the most influential and uncertain among the five driving forces identified by the experts. The scenarios did also incorporate variations of the remaining driving forces (i.e. customers' attitude, economy and environment). The four scenarios are the following: *(1) AV ...in standby, (2) AV...in bloom, (3) AV...in demand, (4) AV... in doubt.*

The first scenario *AV...in stand by* describes a path where fully automated vehicles become available in 2030. The technology develops rapidly but government policies are not supportive because the Dutch government foresees substantial risk and negative impact associated with this new transport technology. Thus, industry takes the lead in the development of automated vehicles. Strongly induced travel demand, sprawling trends and pressure on conventional public transportation services follow the introduction of automated vehicles.

The second scenario *AV...in bloom* describes a path where fully automated vehicles become available in 2025. Both technology and policies of the Dutch government support rapid development of this new transport technology. The positive economic context and other concurrent societal changes of this period (i.e., growth of shared economy, environmental awareness movement) drive the demand for (shared) automated vehicles high. Travel demand management and other regulatory measures are necessary to curb the rapidly increasing vehicle travel in the Netherlands.

The third scenario *AV...in demand* describes a path where fully automated vehicles become available in 2040. Dutch government policies promote development of automated vehicles because it expects major societal benefits (e.g. congestion relief, less accidents) from this technology. Yet, the complexity of urban environment along with the first fatal accidents hinders technological development.

Penetration rate of automated vehicles rises significantly only after 2040. An important increase in travel demand follows during subsequent years as a result of a decrease in value of time and an increase in capacity.

The fourth scenario *AV…in doubt* describes a path where fully automated vehicles become available only in 2045 but remain a marginal mobility technology thereafter, mainly available as a premium service. Slow technological development, lack of supportive policies and skeptical customers form a highly negative context for automated vehicles.

According to this scenario analysis conditionally and fully automated vehicles are expected to hit the market between 2018–2028 and 2025–2045 respectively. The introduction of fully automated vehicles will likely drive demand for this mobility technology higher. However, the complexity of the urban environment and unexpected incidents like fatal accidents; bankruptcy or change in strategic priorities of major industry players could significantly influence the development path of automated vehicles. The introduction of automated vehicle will likely have implications on mobility in all scenarios. It is expected that the Dutch government will need to introduce travel demand management measures in three out of the four scenarios. Finally, experts assessed scenario 2 *AV….in bloom* and scenario 3 *AV … in demand* as the most likely to unfold in the future.

3.1.3 Implications of Scenario Analysis for Planning

Scenarios can assist long-term planning in developing strategies to quickly and effectively adapt to major and unpredictable challenges of the future. Stead and Banister [8] suggest that scenario analysis can provide us with insights to avoid adverse outcomes in the future. Such analysis can also enhance robustness of transport policy and infrastructure investment decisions. However, development of scenarios can also be seen as the first step towards a more rigorous examination of potential impacts of automated vehicles. For example, the results of a scenario study could provide input to agent-based, activity-based and integrated land use—transportation models to simulate potential implications of automated vehicles for travel demand and externalities such as energy consumption and emissions (see e.g. [9]).

3.2 *Pilot Projects*

We need a better understanding supported by data of what will happen on the demand side and supply side of transportation. Therefore, pilot projects on real behavior will be useful. An audience member noted that it is helpful to think of the product development principle that states you do not know how someone will use a product until it is in their hands—not in a test facility, but taken home and used in daily life. Therefore, several pilot projects each focused on a different aspect of behavior would be valuable in collectively providing data to develop policies and models. A key challenge here is that the technology is not fully available to the public and researchers for widespread testing. However, we could also gain insights from data of existing taxi, car sharing and ride sharing services which start mimicking aspects of autonomous vehicles and shared-mobility systems. Finally, another key question is how we can aggregate data at the global level, so that projects around the world can gain insights from each other. The community should also identify some universal performance metrics for evaluating pilot projects.

3.3 Education

We need to educate the planning community, including local and state elected officials. We need to educate the public. However, public education is difficult, and it is difficult to convince legislators to care about something that is far in the future. Some agencies are providing education about new technologies within their agencies, for example Florida Department of Transportation organizes an annual Automated Vehicles Summit. We need to continue the conversation through channels such as webinars, the Association of Metropolitan Planning Organizations (AMPO), Florida meeting, TRB. There are research roadmaps for connected vehicle and automated vehicles prepared by AASHTO and funded through and NCHRP pooled study. Another audience member mentioned the American Association of Motor Vehicle Administrators (AAMVA), as a working group for states and a good resource on licensing. Widespread education is also important to ultimately define the roles and responsibilities in addressing AVs at the local, MPO, state and Federal level.

4 Summary and Conclusions

Despite the considerable uncertainties associated with the capabilities and deployment time lines of the AV technology, it is important for planners to start considering AV technology in their long-range-transportation planning process. This paper first described three major sources of uncertainty as technology, policy and user behavior. The implications of these uncertainties for transportation planning were discussed next. The discussion led to the identification of several key questions that planners have to start thinking about today. The third section of the paper presented "scenario planning" as a systematic procedure to discuss uncertain futures. A case study from the Netherlands was presented as an illustration. It is envisioned that many MPOs will undertake their own scenario planning exercise. The need for pilot projects to understand behavioral implications of new technologies was also emphasized. Finally, the need to educate all involved stakeholders was stressed. All the material presented in this paper is drawn from the presentations and audience discussions at a workshop at the 2015 Automated Vehicles symposium. The workshop organizers plan to continue to engage stakeholders via webinars and presentations/sessions at other conferences.

References

1. Misener J (2015) Implications of vehicle automation for planning, presentation at the breakout session on planning, automated vehicles symposium, Ann Arbor
2. SAE International (2014) Taxonomy and definitions for terms related to on-road motor vehicle automated driving systems. SAE International, Warrendale, PA. http://www.sae.org/misc/pdfs/automated_driving.pdf Accessed 12 Oct 2015
3. Mudge R (2015) Implication of vehicle automation for planning—policy links, presentation at the breakout session on planning, automated vehicles symposium, Ann Arbor
4. Smith S (2015) User issues, automated vehicles symposium, breakout session on planning, Ann Arbor
5. Smith S (2015) User issues, presentation at the breakout session on planning, automated vehicles symposium, Ann Arbor
6. Mahmoud M, Liu Y, Hartmann H, Stewart S, Wagener T, Semmens D, Stewart R, Gupta H, Dominguez D, Dominguez F, Hulse D, Letcher R, Rashleigh B, Smith C, Street R, Ticehurst J, Twery M, van Delden H, Waldick R, White D, Winter L (2009) A formal framework for scenario development in support of environmental decision-making. Environ Model Softw 24 (7):798–808
7. Milakis D, Snelder M, van Arem B, van Wee B, Correia G (2016) Scenarios about development and implications of automated vehicles in the Netherlands. In: Transportation Research board 95th annual meeting. Washington DC, TRB, 2016
8. Stead D, Banister D (2003) Transport policy scenario-building. Transportation Planning and Technology 26(6):513–536
9. Childress S, Nichols B, Coe S (2015) Using an activity-based model to explore possible impacts of automated vehicles. In: Transportation research board 94th annual meeting, Washington DC, TRB, 2015

Printed by Printforce, the Netherlands